5th International Workshop & Summer School on Plasma Physics 2012

Journal of Physics: Conference Series Volume 715

Kiten, Bulgaria
25 – 30 June 2012

Editor:

Evgenia Benova

ISBN: 978-1-5108-2419-5
ISSN: 1742-6588

Printed from e-media with permission by:

Curran Associates, Inc.
57 Morehouse Lane
Red Hook, NY 12571

proceedings.com

Some format issues inherent in the e-media version may also appear in this print version.

Printed by Curran Associates, Inc. (2016)

Institute of Physics
Dirac House, Temple Back
Bristol BS1 6BE UK

Phone: 44 1 17 929 7481
Fax: 44 1 17 920 0979

techtracking@iop.org

Additional copies of this publication are available from:

Curran Associates, Inc.
57 Morehouse Lane
Red Hook, NY 12571 USA
Phone: 845-758-0400
Fax: 845-758-2633
Email: curran@proceedings.com
Web: www.proceedings.com

TABLE OF CONTENTS

Journal of Physics: Conference Series **715** (2016) 011001 doi:10.1088/1742-6596/715/1/011001

EDITORIAL

Fifth International Workshop & Summer School on Plasma Physics 2012

Evgenia Benova 2016 *J. Phys.: Conf. Ser.* **715**

The Fifth International Workshop & Summer School on Plasma Physics (IWSSPP'12) was organized by St. Kliment Ohridsky University of Sofia, with co-organizers TCPA Foundation, Association EURATOM/IRNRE, and the Bulgarian Academy of Sciences.

It was held in Kiten, Bulgaria, on the Black Sea coast, from June 25–30, 2012. The scientific programme covers the topics ***Fusion Plasma and Materials; Plasma Modeling and Fundamentals; Plasma Sources, Diagnostics and Technology***.

The 4th edition of the Workshop ***Plasmas for Environmental Issues*** was co-organized together with the Institute of Plasmas and Nuclear Fusion, Lisbon, Portugal.

A special Workshop on ***Remote GOLEM operation*** was organized by the Institute of Plasma Physics, Prague, Czech Republic for the students and interested participants to work remotely with the Czech TOKAMAK GOLEM.

As in the previous issues of this scientific meeting its aim was to stimulate the development of and support a new generation of young scientists to further advance plasma physics fundamentals and applications, as well as ensuring an interdisciplinary exchange of views and initiate possible collaborations by bringing together scientists from various branches of plasma physics.

This volume of *Journal of Physics: Conference Series* includes 12 papers (invited lectures, contributed talks and posters) devoted to various branches of plasma physics, among them fusion plasma and materials, dc and microwave discharge modelling, transport phenomena in gas discharge plasmas, plasma diagnostics, material processing, plasma-chemistry and technology. Some of them have been presented by internationally known and recognized specialists in their fields; others are MSc or PhD students' first steps in science. In both cases, we believe they will raise readers' interest.

We would like to thank the members of both the International Advisory Committee and the Local Organizing Committee, the participants who sent their manuscripts and passed through the (sometimes heavy and troublesome) refereeing and editing procedure and our referees for their patience and considerable effort to improve the manuscripts. We would like to express our gratitude to the invited lecturers who were willing to pay the participation fee. In this way, in addition to the intellectual support they provided by means of their excellent lectures, they also supported the school financially.

We are grateful to FuseNet (http://www.fusenet.eu/) for supporting European masters' and PhD students' participation in this event, and to the Fund for Scientific Research at the University of Sofia, Grant No 145/2012.

E. Benova

Journal of Physics: Conference Series **715** (2016) 011002 doi:10.1088/1742-6596/715/1/011002

Peer review statement

All papers published in this volume of *Journal of Physics: Conference Series* have been peer reviewed through processes administered by the proceedings Editors. Reviews were conducted by expert referees to the professional and scientific standards expected of a proceedings journal published by IOP Publishing.

Problem-Oriented Simulation Packages and Computational Infrastructure for Numerical Studies of Powerful Gyrotrons

M Damyanova[1], S Sabchevski[1], I Zhelyazkov[2], E Vasileva[1], E Balabanova[1], P Dankov[2], P Malinov[2]

[1]Academician Emil Djakov Institute of Electronics, Bulgarian Academy of Sciences, Association EURATOM-INRNE 72 Tzarigradso chaussee Blvd., BG-1784 Sofia, Bulgaria

[2]Faculty of Physics, Sofia University, Association EURATOM-INRNE 5 James Bourchier Blvd., BG-1164 Sofia, Bulgaria

E-mail: milenadam@gmail.com

Abstract. Powerful gyrotrons are necessary as sources of strong microwaves for electron cyclotron resonance heating (ECRH) and electron cyclotron current drive (ECCD) of magnetically confined plasmas in various reactors (most notably ITER) for controlled thermonuclear fusion. Adequate physical models and efficient problem-oriented software packages are essential tools for numerical studies, analysis, optimization and computer-aided design (CAD) of such high-performance gyrotrons operating in a CW mode and delivering output power of the order of 1–2 MW. In this report we present the current status of our simulation tools (physical models, numerical codes, pre- and post-processing programs, etc.) as well as the computational infrastructure on which they are being developed, maintained and executed.

1. Introduction

The gyrotrons are fast-wave free-electron devices (gyromonotron oscillators) operating on a physical principle known as cyclotron resonance maser (CRM) instability and thus sometimes referred to as CRM. They are among the most powerful sources of coherent radiation working in a CW (continuous wave) mode in the millimetre and the sub-millimeter wavelength regions of the electromagnetic spectrum and have numerous applications in the physical research and in various technologies [1, 2]. Certainly the most notable one, however, is their usage in the systems for electron cyclotron resonance heating (ECRH) and electron cyclotron current drive (ECCD) of magnetically confined plasmas in reactors for controlled thermonuclear fusion (e.g., tokamaks as, for example, ITER and DEMO) [3]. In recent years, the gyrotrons for fusion have demonstrated a remarkable progress reaching megawatt levels of the output power at frequencies of 140–170 GHz in a long pulse operation [1]. Their computer-aided design (CAD) and optimization is based on numerical experiments carried out using problem-oriented software packages for simulation of different subsystems of the tube (e.g., the electron-optical system (EOS), resonant cavity, internal mode converter, and so forth), i.e., applying the decomposition principle. Such computer codes are actively developed worldwide (see, for example [4] and the references therein). In this paper, we present briefly and illustrate the current status of the simulation tools that are being developed and maintained by our research team in the framework of the collaboration with the leading institutions of the European gyrotron consortium (EGYC), IHM-KIT (Germany) and CRPP-EPFL (Switzerland). An outlook for the future work is given as well.

2. Simulation tools for numerical investigation and CAD of gyrotrons

The gyrotrons are complex vacuum tubes that consist of several subsystems (e.g., electron-optical system (EOS) that forms a helical electron beam, in which most of its energy is associated with the rotational motion of gyrating electrons, i.e., "pumped beam" with high pitch factor; electrodynamical system with a resonant cavity where the interaction of the electron beam and the excited high-frequency electromagnetic field takes place; quasi-optical system for formation of well collimated wave beam and its coupling to the output window, etc.). Therefore, following a reductionist approach (decomposition principle) different physical models are used to describe different subsystems. For example, the formation of the electron beam in the EOS is simulated by fully relativistic self-consistent static or time-dependent physical models implemented in ray-tracing (trajectory analysis) or PIC (Particle in Cell) codes, respectively. Analogously, the interaction in the cavity resonator is described by physical models which can be linear, nonlinear, time-dependent, static, and so on. Such big variety of physical models (e.g., full-wave, quasi-optical or based on one or another diffraction theory) is characteristic also for the system responsible for mode conversion, transmission and coupling of the wave beam. All this explains the necessity to use a great number of computer codes, combined into problem-oriented software packages on the basis of their common functional assignment. Figure 1 shows the structure of our simulation tools, i.e., both the main problem-oriented packages and the computational platforms on which they are installed and used.

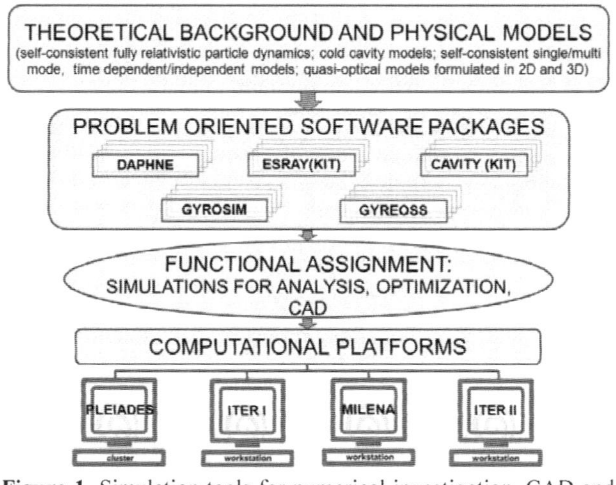

Figure 1. Simulation tools for numerical investigation, CAD and optimization of gyrotrons: problem-oriented software packages and computational infrastructure.

2.1. Current status and functionality of the problem-oriented software packages

Each problem-oriented package contains components (programs, subroutines, and so forth), from which the computational modules are built using appropriate numerical libraries and programming environments (IDE, compilers, debuggers). Some of them have a graphical user interface (GUI) and/or programs for pre-/post-processing and visualization of the results of numerical experiments. Below, these packages are annotated briefly.

2.1.1. DAPHNE

DAPHNE package [4] (developed at CRPP-EPFL) is a programing environment for optimization of EOS of gyrotrons. It is based on an adequate self-consistent physical model (formulated in 2.5D) which consists of a field part (a boundary value problem with Dirichlet and Neumann boundary conditions for the Poisson equation that governs the electrostatic potential distribution taking into account the space charge) and a dynamical part, which contains the relativistic equation of motion of charged macro particles representing the electrons of the beam. The computational region (2D meridional cross-section of an axially symmetric EOS) is discretized using a structured mesh with rectangular cells. DAPHNE is embedded in the ASTRID problem solving environment, which includes: a data base management system for memory and data handling (MEMCOM); 3D adaptive mesh generator; ASTRID finite element solver, graphic system for visualization, command language and interfacing modules. DAPHNE is implemented as a script written in ASTRID's command language and includes the two modules CFI (for calculation of the magnetic field of the coils of the tube) and PART (for integration of the equations of particles motion) written in FORTRAN. The script invokes both successively the particle pusher (PART) and the Poisson solver in an iterative loop until the process converges to a self-consistent solution.

2.1.2. ESRAY (KIT)

ESRAY (KIT) [5] is also a problem-oriented package for trajectory analysis (ray-tracing) of EOS based on a fully relativistic 2.5D electrostatic physical model. Its most characteristic distinguishing features are: (i) object-oriented program implementation in C++; (ii) advanced mesh generator which discretizes the computational domain with a great accuracy by structured boundary-fitted grids; (iii) versatile post-processing capabilities and visualization of all scalar and vector physical fields by colour maps; (iv) fast own solver for the boundary value problem by a finite difference method. The package consists of several modules: GRIDGEN (for geometry description and mesh generation), MAGGEN (for calculation the magnetic field of a system of solenoids), ESRAYS (for iterative solution of the self-consistent field problem), and OVIS. The latter module serves as a GUI and postprocessor that presents and visualizes the results of the simulation.

2.1.3. CAVITY (KIT)

The problem-oriented software package CAVITY (KIT) consists of a hierarchy of codes that begins with simple programs (e.g., for an analysis of the mode spectrum; cold cavity code, single mode self-consistent code) and culminates in the most sophisticated self-consistent multimode time-dependent code SELFT. Both the structure of the package and the physical models implemented in its modules has been reviewed recently [6]. The codes are written in FORTRAN and are invoked through a GUI. The GUI itself is in fact a Tcl/Tk script for a Linux (Unix) bash shell that controls: (i) the interaction with the codes, (ii) the specification of the input data, and (iii) the visualization of the results using a set of single commands in the menu window.

2.1.4. GYROSIM

GYROSIM is a problem-oriented software package which includes numerical libraries and source codes of various computational modules (standalone programs, subroutines, pre-, post-processing, and visualization codes) for solving a variety of problems pertinent to the simulation and CAD of gyrotrons using a rich set of adequate physical models [7]. Unlike other packages described above, it is not, however, specialized to only one subsystem of the gyrotron tube. Rather, the individual components of GYROSIM are designed for simulation of all main subsystems of the gyrotron tube, notably: (i) the electron-optical system (EOS), (ii) the magnetic system which includes the main magnet and an arrangement of additional coils, (iii) the electrodynamical system (resonant cavity), and (iv) the quasi-optical system for mode conversion and transmission of the radiation.

It should be mentioned that the codes for numerical modelling of the EOS (GUN-MIG/CUSP) are based on a 2.5D physical model, which is analogous to the one implemented in DAPHNE and ESRAY and therefore provide results that are consistent and in a good agreement with each other. Besides the differences in their program implementation, GUN-MIG/CUSP, however, allows magnetron injection guns (MIG) with a reversal of the magnetic field (e.g., a magnetic cusp) that form axis-encircling (aka uniaxial) beams to be simulated with an increased accuracy. Similarly, the codes of GYROSIM for simulation of the electrodynamical system cover the same functionality as the CAVITY (KIT). At the same time, there are some notable differences between them. For instance, the CAVITY (KIT) can treat both conventional and coaxial resonators but at fundamental operation while the cavity codes belonging to GYROSIM are specialized only to cavities without an insert but can simulate operation at the second (and in the case of a large orbit gyrotron (LOG) even higher) harmonics of the cyclotron frequency. In its current form, the GYROSIM is a heterogeneous package and includes components written in different languages (Fortran 77, Fortran 90, C, C++, SciLab, and so on), operational and/or portable to different computational platforms (ranging from laptops and workstations to mainframe and supercomputers), and executable under different (genuine as well as emulated/virtualized) operating systems (e.g., Unix, Linux, Windows, Cygwin). Another characteristic feature of the package is that it is being built following a concept of extensibility which allows us to add/replace easily different computational modules and in such a way to modify both the numerical algorithms and the physical models implemented in the programs. The latest upgrade of GYROSIM package has been carried out in parallel with the development of a novel module called GO&ART (which stands for Geometric Optics and Analytic Ray Tracing). It consists of several codes (RAYS, COMODES, and TRACE) for analysis of quasi-optical components (Vlasov and Denisov type launchers, reflectors and phase-correcting mirrors, and so on) as well as systems based on them (e.g., internal mode converters and transmission lines). As an illustration, some screenshots of these programs are shown in figure 2.

2.1.5. GYREOSS

Initially, GYREOSS has been conceived as a package of codes for simulation of EOS using a physical model formulated in three space dimensions in order to take into account the departure from axial symmetry due to various misalignments (for instance of the electrodes, of the magnetic coils, etc.) and non-uniformities. Its initial version has been implemented using the gmsh package for meshing, pre- and post-processing and GetDP as a solver [8]. In the recent years, however, GYREOSS has evolved as a test bench for experimenting with different numerical methods, solvers and algorithms in 3D aiming the final goal – a parallel 3D code for numerical simulation and CAD of EOS of gyrotrons.
The latest version of GYREOSS is being developed using the FreeFEM++ problem solving environment [9]. Recently a series of experiments have been carried out in order to study both the accuracy and the speed of several experimental electrostatic solvers of GYREOS based on different Finite Element Method's (FEM) formulations and various combinations of finite elements (e.g., the Raviart–Thomas finite elements, real-time three-dimensional (RT3d) vector-valued H(div)-conforming finite elements, and continuous piece-wise linear finite elements P03) and a method for solving the resulting set of linear equations. These studies are being complimented by investigations focused on the optimization of the used tetrahedral meshes and their adaptive refinement (AMR) using the Hessian of the solution. At the same time, a preparation for a MPI parallelization of the code is in progress now.

A screenshot that illustrates the GUI and the visualization (based on the medit – an interactive mesh visualization software [10]) used by GYREOSS is shown in figure 3.

2.2. Available computational platforms for code maintenance and development

All packages outlined above (CAVITY, ESRAY, GYROSIM, GYREOSS) are installed and are operational on the workstations of the Bulgarian research team (see figure 1), except DAPHNE, which is available to us for remote execution and maintenance on the PLEIADES2 cluster from Sofia. Since the outstanding performance of PLEIADES2 is well known, we will mention only the basic characteristics of the most powerful of our workstations. ITER I has two CPU AMD Opteron^{TM} Dual Core 275, 2.2 GHz and RAM 4 GB DDRAM with a MB Supermicro-Dual Opteron and SVGA Nvidia

GeForce 6600 TD. The workstation ITER II has 2 CPUs Intel Xeon X5680, 3.33 GHz,12 MB cache, 6 Cores; memory 4×4 GB DDR-31333. On both systems the operating system is Ubuntu 10.04 (lucid), Kernel Linux 2.6.32-41-generic.

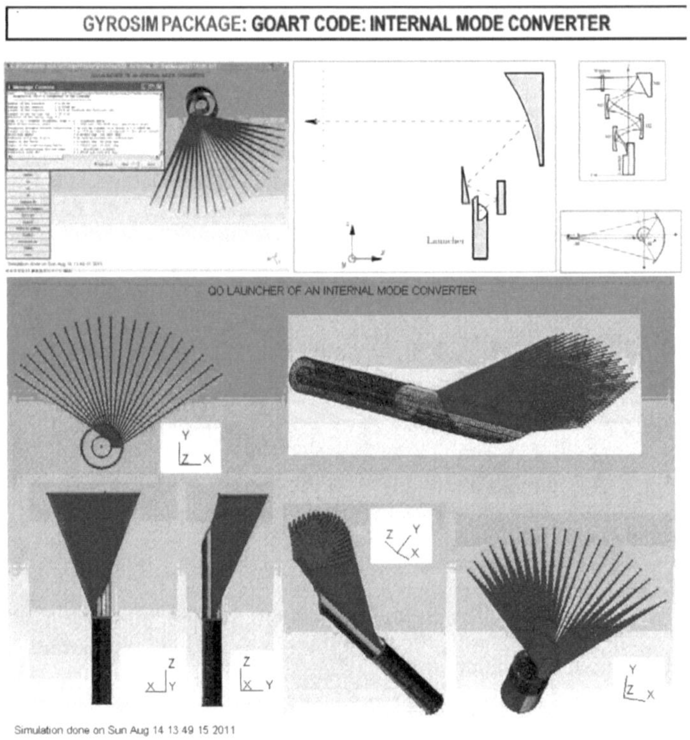

Figure 2. Simulation of quasi-optical components by GO&ART module of the GYROSIM package.

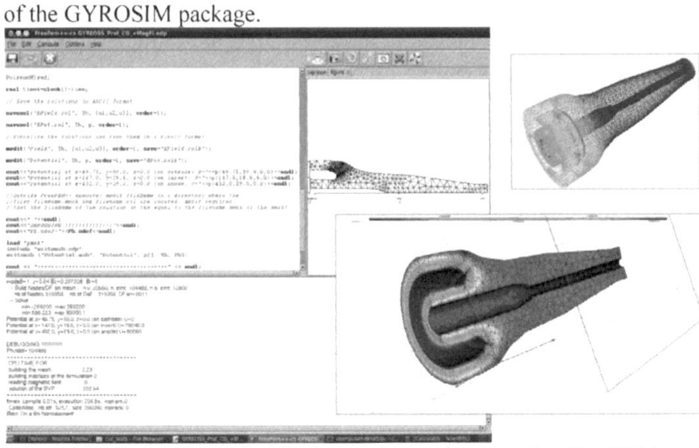

Figure 3. Screenshot of GYREOS obtained using the FreeFEM++-cs GUI. In this example, the EOS of a coaxial 2 MW-170 GHz gyrotron has been simulated.

3. Conclusion

The problem-oriented packages outlined in the previous section are under continuous development and improvement. Recently they have been used in a series of numerical experiments, carried out to study the designs of powerful gyrotrons that are under consideration and/or development at present [5]. The conducted simulations give a deeper physical insight into the operation of high-performance gyrotrons of megawatt class and are good benchmarks that demonstrate the improved capabilities and functionality of the upgraded codes. Moreover, these results suggest some further experiments for more detailed study of the correlation between the beam-quality parameters and efficiency, on one hand, and the particular design (configuration of the electrodes, tailoring of the magnetic field, etc.), on the other hand. It is expected that the novel and upgraded versions of the simulation packages will contribute to the development of the next generation of powerful gyrotrons for fusion with an improved performance.

Acknowledgements

This work was carried out in the framework of Task 2.1.2 of the scientific programme of the Association EURATOM-INRNE in collaboration with the gyrotron teams at IHM-KIT (Karlsruhe, Germany) and CRPP-EPFL (Lausanne, Switzerland).

Reference

[1] Thumm M 2012 State-of-the-art of high power gyro-devices and free electron masers (Update 2011) *KIT Scientific Reports* **7606** 1–114, ISSN 1869-9669

[2] Idehara T, Saito T, Ogawa I, Mitsudo S, Tatematsu Y, Sabchevski S 2008 The potential of the gyrotrons for development of the sub-terahertz and the terahertz frequency range – A review of novel and prospective applications, *Thin Solid Films* **517** 1503–1506

[3] Erckmann V, Kasparek W, Plaum B, Lechte C, Petelin M, Bruschi A, D'Arcangelo O, Bin W, Braune H, van den Braber R, Doelman N, Gantenbein G, Laqua H P, Lubiako L, Marushchenko N B, Michel G, Thumm M 2011 Large Scale CW ECRH Systems: Meeting a Challenge, in: RADIO FREQUENCY POWER IN PLASMAS: Proceedings of the 19th Topical Conference, *AIP Conference Proceedings* **1406** 165–172

[4] Sabchevski S, Zhelyazkov I, Thumm M, Illy S, Piosczyk B, Tran T-M, Hogge J-Ph, Pagonakis J Gr 2007 Recent Evolution of the Simulation Tools for Computer Aided Design of Electron-optical Systems for Powerful Gyrotrons *Computer Modeling in Engineering and Sciences* **20** 203–220

[5] Damyanova M, Balabanova E, Kern S, Illy S, Sabchevski S, Thumm M, Vasileva E, Zhelyazkov I 2012 Simulation tools for computer-aided design and numerical investigations of high-power gyrotrons *J. Phys.: Conference Series* **356** 012044, 4 pages

[6] Damyanova M, Kern S, Illy S, Thumm M, Sabchevski S, Zhelyazkov I and Vasileva E 2012 Modelling and Simulation of Gyrotrons for ITER *Proc. 4th International Workshop & Summer School on Plasma Physics* (Kiten, Bulgaria, July 5–10, 2010), 6 pages – in press

[7] Sabchevski S, Idehara T, Saito T, Tatematsu Y, Ogawa I, Mitsudo S 2010 Physical Models and Computer Codes for CAD and Study of Gyrotrons (Invited talk) *Proc. 3rd International Workshop on Far Infrared Technologies* (IW-FIRT 2010), Paper 15a_4, 8 pages

[8] Damyanova M, Sabchevski S, Zhelyazkov I 2010 Pre- and Post-processing of Data for Simulation of Gyrotrons by the GYREOSS Software Package *J. Phys.: Conference Series* **207** 012032, 6 pages

[9] Hecht F 2010 *FreeFem++*, Third Edition, Version 3.19-1, Laboratoire Jacques-Louis Lions, Université Pierre et Marie Curie, 75252 Paris Cedex 05, France, 1–378. Visit: http://www.freefem.org/ff++/ftp/freefem++doc.pdf

[10] Frey P J 2001 **MEDIT** An interactive mesh visualization software, Institut National de Recherche en Informatique et en Automatique, *Rapport Technique* No 0253 (44 pages) ISSN 0249-0803. Visit: http://www.ann.jussieu.fr/frey/publications/RT-0253.pdf

Metal hydrides studied in gas discharge tube

I. Bozhinova, S. Kolev, Tsv. Popov, A. Pashov
Department of Physics, Sofia University, 5 J. Bourchier blvd., 1164 Sofia, Bulgaria
E-mail: pashov@phys.uni-sofia.bg

M. Dimitrova
Institute of Electronics, Bulgarian Academy of Sciences, Tsarigradsko schausee 72, 1784 Sofia, Bulgaria

Abstract. A novel construction of gas discharge tube has been tested for production of high densities of metal hydrydes. Its performance turned out to be comparable with the existing sources of the same type and even better. First results of the tests on NiH are reported and critically analysed. Plans for future modifiaction of the construction and application of the tube are discussed.

1. Introduction

The metal hydrides (especially FeH, CrH, NiH) are among the most abundant molecules that have been observed in the Sun spots [1, 2, 3] and in the atmosphere of other cold stars [4, 5]. Therefore the knowledge of their absorption spectra is important in order to identify them and this is one of the motivations for laboratory studies of this molecules. Moreover by using the Zeeman splitting of the rovibrational energy levels it is possible to make estimations on the magnetic field on the surface of space objects [6, 7]. The hydrides are specially suited for this, because their rotational constants are large and it is relatively easy to observe spectra with fully resolved rotational structure.

Many papers have been devoted to spectroscopic studies of metal hydrides (MH) (see for example [5, 8] and references therein). There are too main approaches for production of these molecules in laboratory conditions. The first utilizes a King type furnace and the molecules are produced by heating the metal up to 2500 - 3000 K in a H_2 containing atmosphere [9]. The second approach uses gas discharge tubes, where the metal is vaporized by sputtering [10, 11, 12]. The second approach has the advantages of producing molecules at lower gas temperatures (close to room temperature) with smaller Doppler broadening. The gas pressure is also lower (≈ 1 torr compared with several tens torrs in the King furnaces) leading to smaller pressure broadening. On the other hand, the collisions with electrons allow higher rovibrational levels of the ground electronic state to be populated which makes the spectroscopic information more rich.

Some of the existing hollow cathode sources provide relatively high concentration of MH molecules, but in a small volume [12]. The current densities are high which leads to fast sputtering of the cathode and requires a cooling. Another construction of gas discharge source can be found in [10]. It consists of a long tubular cathode and a set of pin anodes placed in the centre of the tube. It overcomes the disadvantages of the hollow cathode source. It has large absorption length and less sputtering of the cathode, but the emmision from the discharge fills the centre of the tube which makes laser-induced fluorescense (LIF) measurements, for example, difficult.

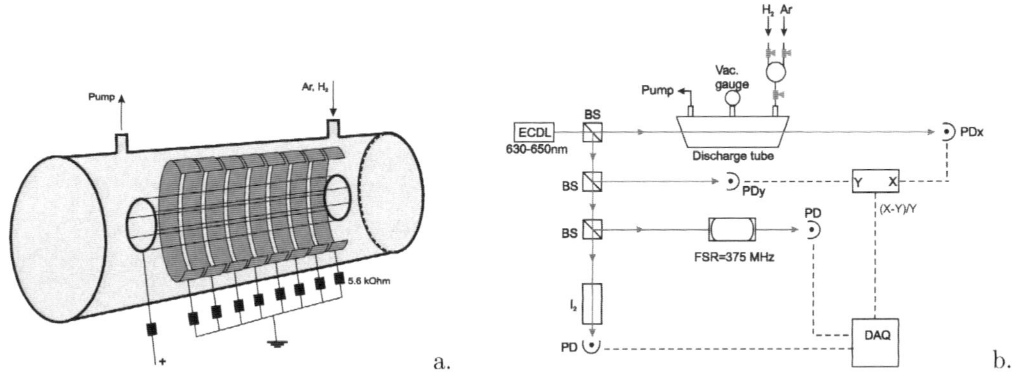

Figure 1. a. Scheme of the discharge tube and **b.** the experimental setup for absorption spectroscopy.

In this paper we present a different design of the discharge tube. It consists of a series of separated ring cathodes and an anode made of several rods forming a cylinder around the center of the tube (Figure 1a). At certain conditions (pressure and discharge current) it was shown that a dark zone is formed in the center of the tube when filled with Ar [13]. This observation motivated us to undertake a systematic study of this tube in order to figure out (i) whether it is suitable for production of high densities of MH and (ii) is it possible to produce molecules at the same conditions when a dark zone is formed in the center of the tube. This would make this tube an extremely attractive source for LIF spectroscopy. The present experiments were carried out with Ni cathodes, because several sources have been tested for NiH production [10, 11, 12] providing good ground for comparison.

2. Experimental setup and results

The discharge tube is made of 50 cm long pyrex glass tube. A total of 8 cathodes (a ring made of Ni 0.5 mm foil, 50 mm in diameter and 20 mm thick) are placed. The space between the rings is about 5-6 mm. Each ring is connected separately to the ground of the high voltage (HV) source via a blast resistor of 5.6 kOhm through a glass feedthrough. The anode consists of six Ni rods, 300 mm in length forming a cylinder with a diameter of 30 mm. The tube is closed with glass windows at a small angle with respect to the tube axes (to avoid back reflections to the laser source and interferences). The tube is equipped with two additional glass tube connections, where KF flanges are mounted for pumping out the tube and filling with buffer gas. The DC HV source is able to reach 1.5 kV and currents up to 250 mA. At these conditions and gas pressures down to 100 mtorr of H_2 and Ar we observed a very stable and homogenous discharge between the cathodes and the anodes.

The concentration of NiH molecules was monitored through their absorption in the red $B^2\Delta_{5/2}$-$X^2\Delta_{5/2}$ band starting from $v''=0$ in the ground state to $v'=0$ in the excited electronic B state. This band has been observed already by Brien et al. [11]. The authors give a list of frequencies of identified transitions in two isotopologues – ^{58}NiH and ^{60}NiH so it should be easy to set the laser to one of the absorption lines. The setup is shown in Figure 1b. The laser source is a commercial free running diode laser (635 nm or 650 nm) scanned in frequency by the current and the temperature. While the main laser beam passes the discharge tube and its intensity is measured by the photodiode PDx, three other beams are split from it. The first one goes to PDy for calibration of the laser power. The second and the third are used for frequency calibration of the laser frequency by a confocal interferometer (free spectral range of 375 MHz) and a 300 mm long iodine cell at room temperature. All signals are collected and recorded by a Tektronix TDS 2024 digital oscilloscope.

After setting the laser frequency on a selected line from [11] the laser was scanned by ±10 GHz around it. At the same time various parameters of the discharge (gas composition, current, pressure) were changed searching for absorption signal. Initially we started with a gas mixture of 90 % Ar and 10 % H_2. This concentration was reported to be optimal in Ref. [10, 12]. The experiments have shown, however, that for the current discharge tube, NiH signals are obtained only in pure hydrogen atmosphere. Adding even small amounts of Ar destroys the absorption signals.

In Figure 2a a typical set of signals is shown. Along with the interference fringes and the iodine lines two NiH lines are seen in this scan (J″=8.5) belonging to different istopologues ^{58}NiH and ^{60}NiH. The NiH signals were very weak, corresponding to an absorbance of about 0.02 % (the strongest signals) and less. An estimate of the Doppler width is made $\Delta\nu_D = 0.027$ cm^{-1}, corresponding to gas temperatures of about 340 K, which sounds very reasonable.

From the measured absorption coefficient and the known absorption length L=250 mm the concentration of NiH molecules in a particular rovibrational level N_J was done through the following realtions:

$$\Delta I/I = \sigma(\nu_0) N_J L, \qquad \text{where} \qquad \sigma(\nu_0) = \frac{2\lambda^2 A_{00}}{8\pi\Delta\nu_D}\sqrt{\frac{\ln 2}{\pi}}\,. \tag{1}$$

Here λ is the wavelength and A_{00} is the Einstein coefficient for spontaneous emission from the upper level (v'=0, $J' = 7.5$) to the lower level (v''=0, $J'' = 8.5$) of the electronic transition. We estimated this coefficient as follows. From Ref. [10] it is known that the lifetime of the rovibrational levels with $v' = 1$ of the $B^2\Delta_{5/2}$ state is about 1 μs. From [14] it is known that the overall LIF from the $B^2\Delta_{5/2}(v' = 1)$ levels is about 20 times more intensive than the fluorescence from $B^2\Delta_{5/2}(v' = 0)$. If we assume the induced fluorescense as a measure of the number of the excited molecules, we can estimate the transition ($v'' = 0 \rightarrow v' = 0$) as 20 times weaker than the ($v'' = 0 \rightarrow v' = 1$) one and consequently $A_{00} = A_{01}/20$.

Under this assumptions we can estimate the density of NiH molecules in the ($v'' = 0, J'' = 8.5$) state from eq. (1). We obtain a maximum concentration of $N_J \approx 10^{11}$ cm^{-3} for $J = 8.5$ and $N \approx 10^{12}$ cm^{-3} for the total density of molecules where a thermal distribution over the rovibrational levels is assumed (T=340 K). Although very rough this estimation shows the better performance of this source comparaed with other gas discharge sources, e.g. in Ref. [10] where the authors report a concentration of $N \approx 10^{10}$ cm^{-3}.

In order to optimize the working parameters of the tube, we measured the relative absorption of the same spectral line at different currents and pressures of Hydrogen. The measurements were carried out in the center of the tube and close to the anodes. While the increase of the absorption with the current is very close to linear, there is an optimal gas pressure at about 200 – 400 mtorr of hydrogen (see Fig. 2b). Within the uncertainty of our measurements we did not observe significant change of the absorption close to the anodes, so the distribution of NiH within the anode volume should be pretty homogenous.

3. Discussion

The presented configuration of discharge tube turned out to be a successful alternative of the existing sources for production of metal hydrides. The construction is not limited at length, so long absorption paths can be realized. Additional attractive feature of this tube is that there is no

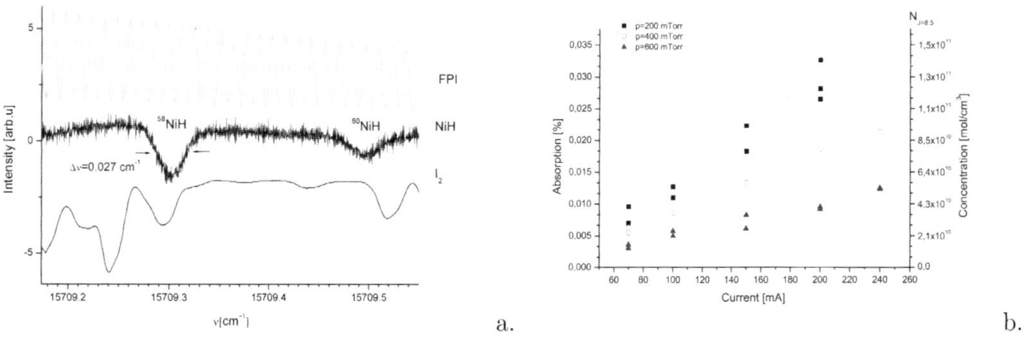

Figure 2. a. Typical absorption spectrum around 15709 cm^{-1}. This is the P(8.5) line in ^{58}NiH and ^{60}NiH). **b.** Dependence of the relative absorption of NiH in the center of the tube on the Hydrogen pressure.

current in its center and at certain conditions it can be dark. Unfortunately in case of hydrogen at the currents and gas pressures for maximum NiH production the central zone was bright. In parallel to the experiment we run a numerical simulation (a report will be presented elsewhere) which reproduces reasonably the experimental observations. At present we are optimizing the code and we will search for configuration with at least similar MH concentrations and dark central zone. One way is to study discharges in different hydrogen containing gases (for example CH_4, H_2O, NH_4). Another possibility is to change the geometry of the tube. Preliminary calculations and also practical consideration have shown that a dark zone can be achieved even at 200-400 mtorr of H_2 when the diameter of the cathodes is about two times larger. There is still another direction for future application of the new tube. We plan to study the possibility to insert the tube in a magnetic field in order to study the Zeeman splitting of the MH lines (especially FeH).

4. Acknowledgements

I.B. and A.P. acknowledge a partial support through the Sofia University grant 143/2012 and the National Scientific Fund of Bulgaria grant RILA 01/1/2011.

References

[1] D L Lambert and E A Mallia 1971 *Mon. Not. R. astr. Soc.* **151** 437–447
[2] H Wohl 1971 *Solar Physics* **16** 362–372
[3] P K Carrol, P McCormackl 1972 *The Astrophysical Journal* **177** L33–L36
[4] John M Brown, Stuart P Beaton, Kenneth M Evenson 1993 *The Astrophysical Journal* **414** L125–L127
[5] Peter Bernath 2005 *AIP Conf. Proc.* **855** 143–148
[6] Jeremy J Harrison, John M Brown, Jinhai Chen, Timothy C Steimle, Trevor J Sears 2008 *The Astrophysical Journal* **679** 854–861
[7] S V Berdyugina and S K Solanki 2002 *A&A* **385** 701–715
[8] T Nelis, S P Beaton, K M Evenson, J M Brown 1991 *Journal of Molecular Spectroscopy* **148** 462–478
[9] S Adakkai Kadavathu, R Scullman, J A Gray, Mingguang Li, R W Field 1990 *Journal of Molecular Spectroscopy* **140** 126–140
[10] M Li, J A Gray, R W Field 1989 *Chemical Physics* **117** 171–176
[11] L C O'Brien, JJ O'Brien 2005 *The Astrophysical Journal* **621** 554–556
[12] R Vallon, S H Ashworth, P Crozet, R W Field, D Forthomme, H Harker, C Richard, A J Ross 2009 *J. Phys. Chem. A* **113** 13159–13166
[13] Popov Tsv, Dimitrova M, Dias F 2004 *Vacuum* **76** 417–420
[14] R Vallon, C Richard, P Crozet, G Wannous, A J Ross 2009 *The Astrophysical Journal* **696** 172–175 and Dr. P. Crozet private communication

Mass Spectrometry of Atmospheric Pressure Surface Wave Discharges

M A Ridenti[1,2], J A Souza-Corrêa[1], J Amorim[1,2]

[1]Laboratório Nacional de Ciência e Tecnologia do Bioetanol - CTBE/CNPEM, P.O. Box 6170, Campinas, S˜ao Paulo, Brazil, Zip Code: 13083-970

[2]Instituto de Fısica Gleb Wataghin, Universidade Estadual de Campinas UNICAMP, Campinas, S˜ao Paulo, Brazil, Zip Code: 13083-859

E-mail: marco.ridenti@bioetanol.org.br, jayr.amorim@bioetanol.org.br

Abstract. By applying mass spectrometry techniques, we carried out measurements of ionic mass spectrum and their energy distribution in order to investigate an atmospheric argon discharge by using a surfatron surface-wave device. The mass and energy distribution measurements were performed with fixed flow rate (2.5 SLM) of pure argon gas (99,999%) and different Ar-O_2 gas mixture compositions (99-1, 98-2 and 97-3). The mass spectra and energy distributions were recorded for Ar^+, O^+, O^+_2, N^+ and N_2^+. The axial distribution profiles of ionic mass and their energy were obtained for different experimental conditions as a function of the plasma length. The results showed that the peak of the positive ion energy distributions shifted to higher energies and also that the distribution width increased as the distance between the sampling orifice and the launcher gap was increased. It was also found that under certain experimental conditions the ion flux of atomic species were higher than the ion flux of their diatomic counterpart. The motivation of this study was to obtain a better understanding of a surface wave discharge in atmospheric pressure that may play a key role on new second generation biofuel technologies.

1. Introduction

Atmospheric plasmas have found many applications in a wide range of fields such as materials processing [1] and biomedical [2]. There are some recent investigations showing that discharges at atmospheric pressure could also be used to process biomass [3, 4]. By applying mass spectrometry techniques [5, 6], we carried out measurements of the mass spectra and the ion energy distribution function of some representative positive species present in a surface wave discharge produced in Ar and Ar-O2 mixtures. Since oxygen containing discharges is a possible candidate to be used in biomass treatment due to the presence of reactive oxygen species, knowing the effect of adding O2 was also important.

The information concerning the ionic species is crucial for the understanding of degradation mechanisms of biomass by ionic impact dissociation in cold plasmas. Positive ions may play an important role on the process of cellulose dissociation and hydrolysis [7]. Moreover, these data can also be used to validate computational models describing the plasma chemistry.

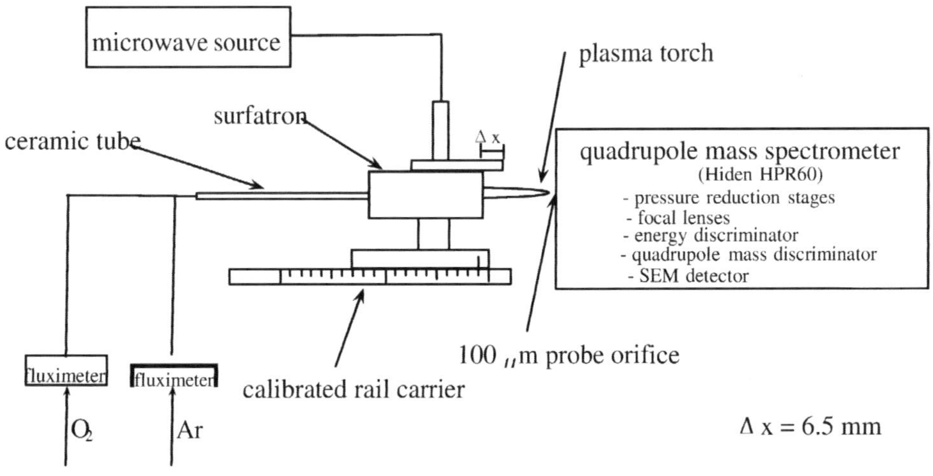

Figure 1. Experimental setup scheme.

2. Experimental Setup

A surfatron surface-wave launcher was connected to a 2.45 GHz Sairem power supply. The gas flowed continuously through a ceramic tube placed inside the surfatron cavity at a flow rate fixed at 2.5 SLM. The flow rate of each component of the mixture was controlled by a MKS flow meter. Measurements were carried out with Ar gas with purity of 99.999% and also with different Ar-O2 mixtures (99:1, 98:2 and 97:3) flowing at a fixed rate of 2.5 SLM. Measurements were performed with applied power of 100 W at atmospheric pressure and room temperature.

The energy distributions were measured by a Hiden EQP mass/energy quadrupole analyzer HPR60. The surface-wave launcher was mounted on an optical rail carrier, which provided displacements in parallel and perpendicular directions on the spectrometer probe plane. The plasma column was aligned to the sampling orifice by adjusting the surfatron position. The diameter of the sampling orifice was equal to 100 μm. The origin of the axial measurements was located at the position where the surfatron was closest to the sampling orifice. That position was 6.5 mm distant from the launcher gap as shown in figure 1. The measurements were performed in 1 mm step along the plasma and the distribution profiles of the ion energy distribution were obtained. A simplified scheme of the experimental setup is showed in figure 1.

3. Ion Energy Distribution Function (IEDF)

When the mass spectrometer is used to measure the positive ion mass spectra the energy of the particles that will be admitted into the quadrupole stage must be selected. In order to find this optimal energy, the ion energy distribution function (IEDF) must be measured at a fixed mass. Figure 2 shows the IEDF of the species whose atomic mass to charge ratio were equal to 40 au/*e*. In this case, these species were the ions Ar^+. This IEDF was measured when the power applied to the plasma source was 100W and a 99:1 Ar-O_2 mixture was used.

The mean energy of the distribution is nearly constant if the distance is lower than 3 mm, but it changes when the distance is higher or equal to 4 mm. Because of that, there is no fixed energy that could simultaneously yield the maximum count rate in all cases. Therefore the total positive ion flux will be better estimated as the IEDF summation,

$$Y_\alpha^a = \sum_i F(\varepsilon_i) \tag{1}$$

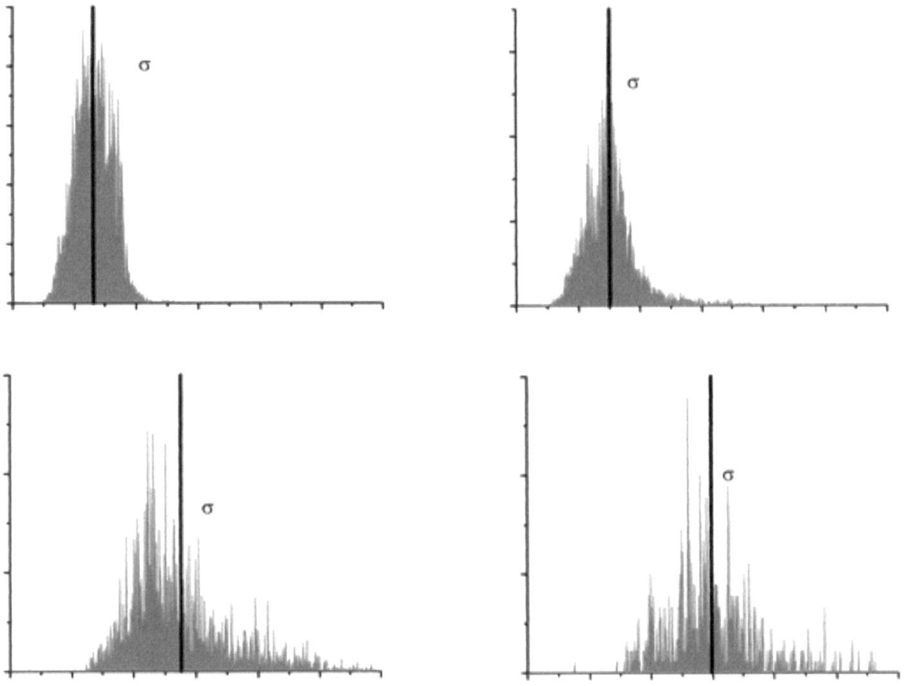

Figure 2. IEDF of the ionic species whose atomic mass to charge ratio were equal to 40 au/e. In this case, these species are the ions Ar^+. This IEDF was measured when the power applied to the plasma source was 100W and a 99:1 Ar-O_2 mixture was used. The IEDF mean energy hEi and standard deviation σ values are also shown. The thick vertical line indicates the mean energy position.

where Y_α^a is the total count rate of specie α and $F_\alpha(\varepsilon_i)$ is the count rate of specie α at energy interval $\varepsilon_i + \Delta\varepsilon$. The relative ion flux Y_α^r of specie α was calculated as the ratio between Y_α^a and the sum of all total fluxes,

$$Y_\alpha^r = \frac{Y_\alpha^a}{\sum_\beta Y_\beta^a} \tag{2}$$

Bruggeman *et al* [6] also reported high mean energies in the IEDF measured in a plasma at atmospheric pressure using the same mass spectrometer and a sampling orifice of 100 μm. They claimed that the IEDF mean energy decreased to values comparable to the thermal energy when a 20 μm sampling orifice was used instead. Indeed, if the plasma sheath length was approximately equal to the diameter of the sampling orifice, the positive ions experienced a nearly colissionless electrode fall. That could also explain the high mean energies of the IEDF we measured. We will assume that in spite of this effect there were no significant changes in the relative abundance of different species.

4. Mass Measurements of Positive Ions

Figures 3 to 6 show the count rates and relative ion fluxes of Ar^+, O^+, O_2^+, N^+ and N^+_2 as a function of distance. The argon (Ar) and oxygen (O2) gas flow rates were controlled in order to obtain relative concentrations of 99:1, 98:2 and 97:3 in the Ar-O_2 mixture. The total flow rate was kept constant at 2.5 SLM.

The result of the measurement in which the discharge was produced in pure Ar is shown in figure 3. Near the gap, at x = 0 mm, the relative ion flux of Ar^+ was much larger than the ion flux of other species as expected, since this plasma region has lower concentrations of species coming from the surrounding air. As the distance increases, the relative relevance of O^+, O^+_2, N^+ and N^+_2 increased, since these ions were originated by species from air diffusing into the discharge and whose concentration must increase as a function of distance. On the other hand, the absolute count rate of all species decreased after a given distance because the plasma degree of ionization and temperature decrease with distance. The absolute count rates of O^+, O_2^+, N^+ and N^+_2 exhibited a maximum value due to the effect of air species diffusing into the plasma.

Another interesting feature of this discharge was that the ratio between the molecular ion and the respective atomic ion decreased as a function of distance. This may be explained by the fact that at the region near the gap heavy species and electrons are hotter, producing high rates of thermal and electron impact dissociation. Since the gas and electron temperature decreases with distance, ion-electron and ion-ion recombination rate of atomic species become much higher than dissociation rates.

When the oxygen (O_2) gas was added to argon, the relative ion flux of O_2^+ and O^+ at point zero were higher than the relative ion flux of N^+_2 and N^+, as expected. In all cases in which a Ar-O_2 mixture was used, the concentration of the atomic ions was often higher than the concentration of the respective molecular ions at points near the gap. At a given point more distant from the gap, however, this relation necessarily changed.

This could also be explained by the fact that at the region near the gap heavy species and electrons are hotter. Nevertheless, only a detailed study of the electron and reaction kinetics

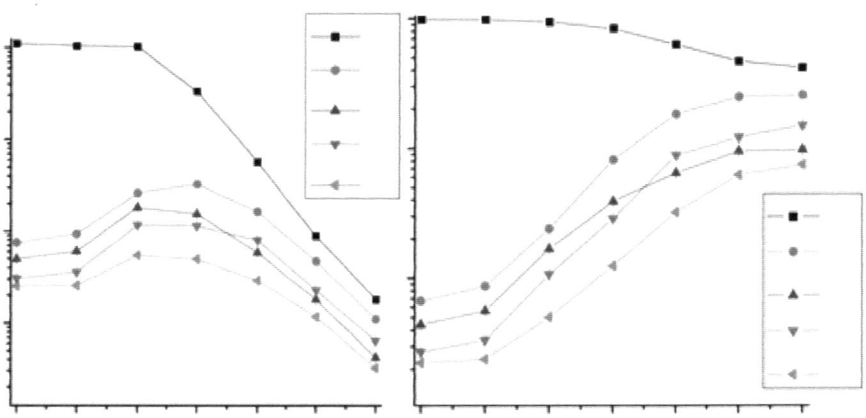

Figure 3. On the left, absolute count rate of Ar^+, O^+, O^+_2, N^+ and N^+_2 as a function of distance. On the right, relative ion flux of Ar^+, O^+, O^+_2, N^+ and N^+_2 as a function of distance. Measurements were carried out with applied power of 100W and pure Ar (99,999%).

Another interesting feature of this discharge was that the ratio between the molecular ion and the respective atomic ion decreased as a function of distance. This may be explained by the fact that at the region near the gap heavy species and electrons are hotter, producing high rates of thermal and electron impact dissociation. Since the gas and electron temperature decreases with distance, ion-electron and ion-ion recombination rate of atomic species become much higher than dissociation rates.

When the oxygen (O_2) gas was added to argon, the relative ion flux of O_2^+ and O^+ at point zero were higher than the relative ion flux of N^+_2 and N^+, as expected. In all cases in which a Ar-O_2 mixture was used, the concentration of the atomic ions was often higher than the concentration of the respective molecular ions at points near the gap. At a given point more distant from the gap, however, this relation necessarily changed.

This could also be explained by the fact that at the region near the gap heavy species and electrons are hotter. Nevertheless, only a detailed study of the electron and reaction kinetics

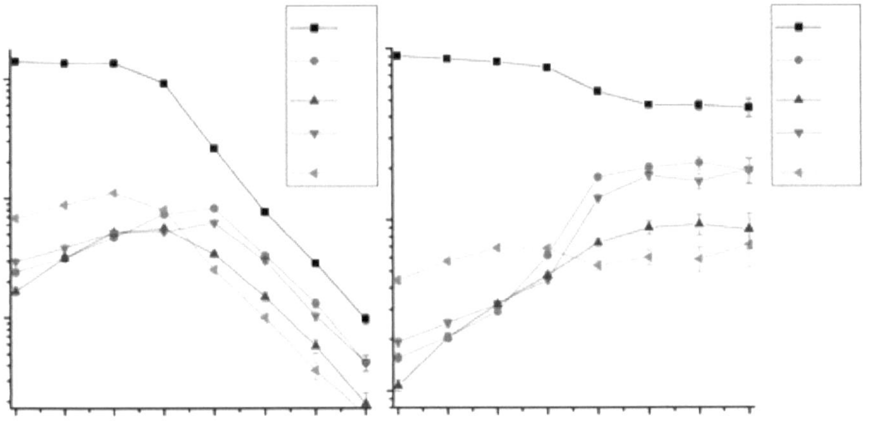

Figure 4. On the left, absolute count rate of Ar^+, O^+, O^+_2, N^+ and N^+_2 as a function of distance. On the right, relative ion flux of Ar^+, O^+, O^+_2, N^+ and N^+_2 as a function of distance. Measurements were carried out with applied power of 100 W and a Ar-O_2 (99:1) mixture.

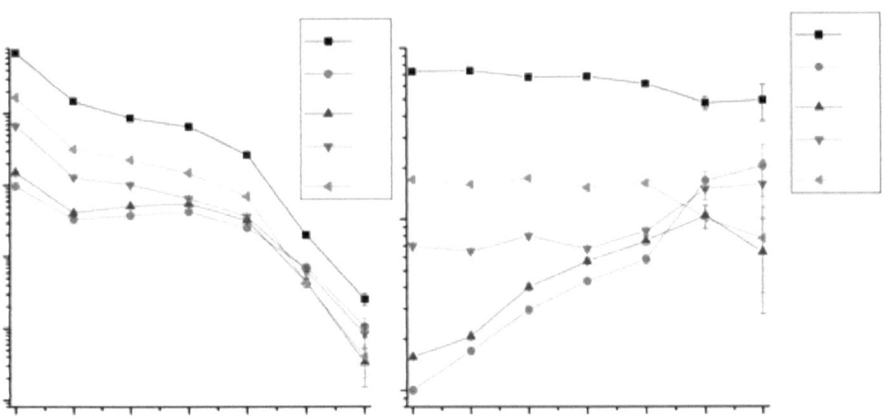

Figure 5. On the left, absolute count rate of Ar^+, O^+, O^+_2, N^+ and N^+_2 as a function of distance. On the right, relative ion flux of Ar^+, O^+, O^+_2, N^+ and N^+_2 as a function of distance. Measurements were carried out with applied power of 100 W and a Ar-O_2 (98:2) mixture.

supplemented with experimental information about electron and gas temperature could explain why most of the molecular positive species were dissociated into its atomic counterpart.

The relative ion fluxes of atomic ions were higher near the gap in all cases, but this effect was more pronounced when gas mixtures 98:2 were used. In the case of the mixture 97:3 and the pair N^+_2 and N^+ this was certainly true only at the distance $x = 0$ mm. For the pair O^+_2 and O^+ the atomic ion flux was higher only if the distance was lower or equal to $x = 3$ mm. On the other hand, when the gas mixture 98:2 was used the N^+ flux was higher from the origin to $x = 3$ mm and the O^+ flux was higher from the origin to $x = 4$ mm. Taking into account the fact that O_2 causes electron attachment, it is expected a drop on electron density as O_2 concentration increases, causing a temperature drop.

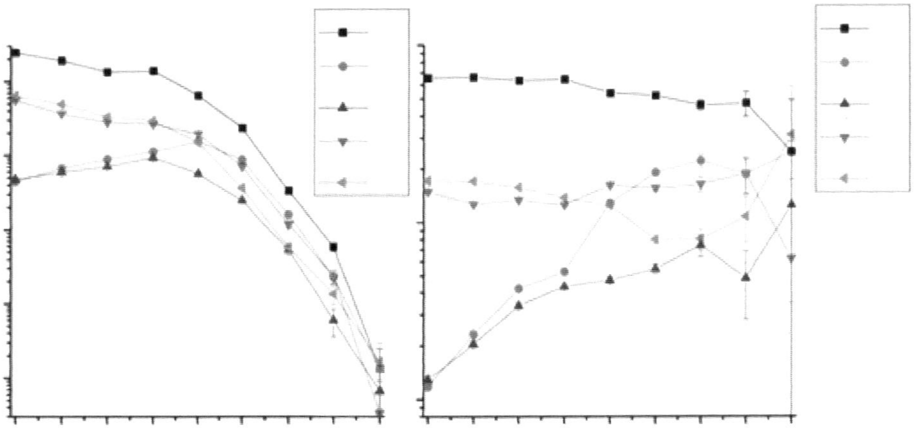

Figure 6. On the left, absolute count rate of Ar^+, O^+, O^+_2 , N^+ and N^+_2 as a function of distance. On the right, relative ion flux of Ar^+, O^+, O^+_2, N^+ and N^+_2 as a function of distance. Measurements were carried out with applied power of 100 W and a Ar-O_2 (97:3) mixture.

This explains why dissociation decreases if O_2 concentration is higher than a threshold value. In order to satisfactorily explain why the mixture 98:2 produced higher dissociation, the electron and the reaction kinetics must be studied. Experimental information about electron and gas temperature would also be useful.

5. Conclusions

The axial distribution profiles of ionic showed that the addition of a small percentage of oxygen (O_2) in the gas mixture changed the relative abundance of oxygen ionic species in the plasma torch. In the region of the plasma column near the launcher gap, not only the oxygen species were dominant in comparison with nitrogen species, but also atomic species were dominant in comparison with their diatomic counterpart. Nevertheless, in the plasma region near the end of the plasma column, diffusion effects always lead to an increase of the concentration of nitrogen species relative to oxygen species. Diffusion effects also explains why the ion flux of nitrogen species first increase before decreasing due to recombination at regions near the end of the plasma column. It was also observed that as the axial coordinate increased, the relative flux of the atomic ions decreased in comparison with the relative flux of their diatomic molecule counterpart, showing that dissociation rates of molecules decreased, probably due to plasma cooling and decreased plasma density.

References

[1] Benedikt J, F F, Yanguas-Gil A and von Keudell A 2006 *Appl. Phys. Lett.* **89** 251504
[2] Stoffels E, Flikweert A J, Stoffels W W and Kroesen G M W 2002 *Plasma Sources Sci. Technol.* **11** 383
[3] Wang C X and Qiu Y P 2007 *Surf. Coat. Technol.* **201** 6273
[4] Souza-Corr^ea J A, Pimenta M T B, Rocha G J M, Gomez E O, Squina F M, Curvelo A A S and Amorim J
[5] 2010 *Plasmas for environmental issues* (Sofia: St. Kliment Ochridsky University) p 68
[6] Malovi'c G, Pua˘c N, Lazovi'c S and Petrovi'c Z 2010 *Plasma Sources Sci. Technology* **19** 034014
[7] Bruggeman P, Iza F, Lauwers D and Gonzalvo Y A 2010 *Plasma Sources Sci. Technol.* **43** 012003 (6 pp)
[8] Song C L, Zhang Z T, Chen W Y and Liu C 2009 *IEEE Transactions on Plasma Science* **37** 1817–1824

A simple method for experimental determination of electron temperature and electron density in a nanosecond pulsed longitudinal discharge used for excitation of high-power atomic and ionic metal and metal halide vapour lasers

K A Temelkov, N K Vuchkov

Metal Vapour Lasers Laboratory, Institute of Solid State Physics, Bulgarian Academy of Sciences, 72 Tzarigradsko Chaussee, 1784 Sofia, Bulgaria

E-mail: temelkov@issp.bas.bg

Abstract. A simple method based on the time-resolved measurement of electrical discharge parameters, such as tube voltage and discharge current, is developed and applied for determination of electron temperature and electron density in the discharge period of a nanosecond pulsed longitudinal discharge, exciting high-power DUV Cu^+ Ne-CuBr, He-Hg^+ and He-Sr^+ lasers.

1. Introduction

Electron temperature and electron density together with gas temperature are the most fundamental plasma parameters in gas discharges and their experimental or theoretical determination is important and relevant problem for understanding and prediction of numerous phenomena in gaseous discharges, laser physics, plasma technologies, gas-discharge mass spectroscopy, absorption and emission spectroscopy, and plasma in general. Electron temperature and electron density determine thoroughly the characteristics constants for elastic and inelastic electron-atom and electron-ion collisions, as well as three-body electron-ion recombination [1-3]. For gas discharge lasers in particular, they directly or indirectly influence on the creation of the inverse population and hence the output laser characteristics.

The strontium ion (Sr^+) recombination laser is a source of violet radiation (430.5 nm and 416.2 nm) with average output power exceeding 1 W [4, 5], and as such, it has potential for application in various areas of lithography, photochemistry and photobiology. The Sr^+ laser is valuable as a pump source for narrow-linewidth blue-green dye lasers. This laser is an ideal violet companion to copper vapour (510.6 nm green and 578.2 nm yellow) and gold vapour (628.3 nm red) lasers for a number of applications. Water-cooled Sr^+ recombination laser, stable-operating at average output power of 1.1 W with independent production of Sr vapour, was developed and studied [4, 5]. The highest average laser power of 1.5 W was achieved under transient conditions. The obtained results were the highest ones for cylindrical geometry of the discharge tube.

A Hg ion (Hg^+) laser, excited in He-Hg nanosecond pulsed longitudinal discharge (NPLD) and stable operating on the 614.9-nm laser transition at average output power of 240 mW, which was the highest one for the gas discharge Hg^+ lasers, was reported in [6, 7]. This laser has potential for application in various areas of photobiology, photochemistry, and spectroscopy. In addition, laser radiation at 614.9 nm could be a valuable instrument at the treatment of the cancer by using photodynamic therapeutics. Until now in this area gold vapour lasers has been successfully used, but their laser tube design and therefore their production are much more complicated.

Laser oscillation was obtained on four deep ultraviolet Cu ion (DUV Cu^+) transitions - 248.6 nm, 252.9 nm, 260.0 nm, and 270.3 nm, in Ne-CuBr NPLD [7, 8]. The small-signal gain of 19 %/m and 16 %/m measured on 248.6 nm and 270.3 nm laser transitions, respectively, was the highest for DUV Cu^+ lasers. An active zone diameter scanning of a DUV Cu^+ Ne-CuBr laser was made [9]. The discharge conditions for achieving of a maximal average output power at the DUV lines were found for each active zone diameter. An average output power of 1.3 W and a specific average laser power of 56.6 mW/cm^3 (at an active volume of 23 cm^3), were obtained at multiline output. The average output power of about 0.85 W and the peak pulse power of about 3.25 W on the 248.6 nm laser line were also the highest for the DUV Cu^+ lasers.

In view of the enormous prospect for application of these lasers, a detailed study on basic plasma parameters, namely gas and electron temperatures and electron concentration is undertaken.

Typical methods to measure the electron temperature include the use of a Langmuir probe and laser Thomson scattering, which are inapplicable to a NPLD. In the literature there are several models, which predict, among the other parameters, values of electron temperature with considerable variation, and furthermore there is no overlap.

Average and time-resolved electron temperatures, which were experimentally determined by the line-ratio method of optical emission spectroscopy, were reported in [10-12] for several metal halide vapour lasers excited by NPLDs. Average values of the electron temperature were also found by averaging the time-resolved electron temperature over the time. Unfortunately, spectral investigation through devices equipped with photomultiplier are prohibited for the discharge period of NPLDs, because of the noise from high-power electrical pulses and the steep rise of voltage and current pulses of $TV.s^{-1}$ and $GA.s^{-1}$.

2. Experimental setup

Discharge zone diameter, which is optimal for laser oscillation on the corresponding transition, and discharge zone length are shown in table 1. Optimal buffer-gas pressure is also presented in table 1.

Table 1. d_a – optimal discharge zone diameter for laser oscillation, l_a – discharge zone length, p – optimal buffer-gas pressure for laser oscillation.

Laser	d_a (cm)	la (cm)	p (Torr)
$He\text{-}Hg^+$	1.5	70	7
$Ne\text{-}Cu^+$	0.71	100	16.7
$He\text{-}Sr^+$	1.6	50	230 / 300

The investigated NPLDs discharges are excited by an electrical scheme with interacting circuits (IC scheme), shown in figure 1 and described in details in [3]. The switch T is a hydrogen thyratron TGI 1000/25. The capacitors (C_1,...,C_4), forming a capacitor bank (CB) are of the low-inductance KVI-3 type. Their capacitance is experimentally optimized for achieving of maximal output laser parameters. The capacitor bank is charged from a high voltage (HV) rectifier. The inductor L_1 is used to level the potential of the electrodes during the period of charging of the CB. The metal plates, P_1 and P_2, with semi cylindrical shape are made of duralumin sheet and are located close to the laser tubes. The pulsed excitation scheme could deliver voltage and current pulses with an amplitude up to 25 kV and 500 A, respectively, an excitation pulse duration of 50–300 ns, and a pulse repetition frequency of 5–30 kHz.

Varying the four capacitors, as well as the matching inductor L, the amplitude, waveform and duration of the pulses of the discharge current and the tube voltage, as well as the phase shift between them, could be manageably changed and controlled. In this way, appropriate discharge conditions for different excitation processes, such as electron impact excitation, charge transfer, Penning ionization, and etc., were found and hence output parameters of various metal atom and ion lasers were increased from 35 % up to two times in comparison with the frequently used electrical pulsed excitation schemes [3, 13].

The waveforms of the discharge current and the tube voltage are detected by a Rogowski-coil probe and a low-inductance Tektronix P6015 HV probe, respectively. All pulses are displayed on a Tektronix 2455A oscilloscope with a 20 MHz cutoff filter.

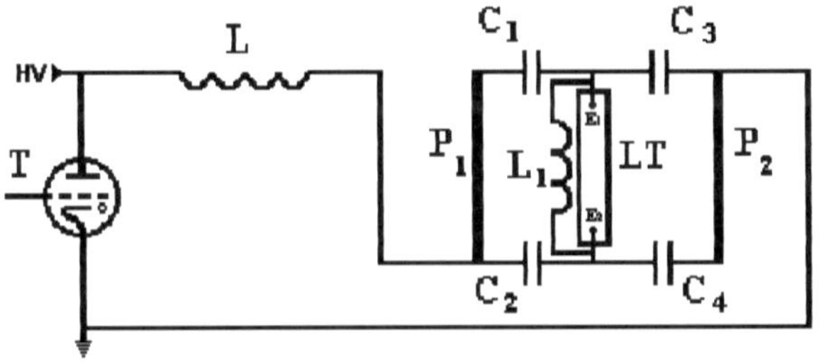

Figure 1. Schematic diagram of the excitation circuit: C1, C2, C3 and C4 - capacitors; L and L1 - inductors; P1 and P2 - metal plates; LT - laser tube; T - thyratron.

3. Experimental results and discussion

As has been done in [3, 14], experimentally determined pulses of the tube voltage U, the discharge current I_R, and the current I_{C3} of the capacitor C_3 are mathematically processed, using the well-known equation (1), in order to eliminate the inductive component of the voltage and to calculate the active voltage U_R, the pulse power $W = U_R.I_R$, and the plasma resistance $R = U_R / I_R$:

$$U_R = U - L_P \frac{dI_{C3}}{dt} - (L_P + L_R)\frac{dI_R}{dt} \qquad (1)$$

where L_R and L_P are the stray inductances of the discharge tube and the interacting circuit, which includes capacitor C_3, respectively, and are calculated from the oscillograms.

The NPLD electric field is uniform, and is determined by the external applied voltage [15]. The use of an electron energy distribution function (EEDF) of Maxwellian form instead of the true EEDF allows good agreement with experiment [15]. It is well known that the electron temperature T_e is a function of the reduced longitudinal electric field E_z / p, which is not a strictly linear function. Experimental data for the function $T_e = f(E_z/p)$ given in [16] for self-sustained PC gas discharge in He, Ne, and Ar allow us to obtain T_e. From the plasma resistance R and the geometrical sizes of the NPLDs (see table 1) together with data presented in table 2 [17] electron density n_{e1} can be estimated.

Table 2. Estimated values of electron mobility μ_e, effective collisions frequency for momentum length ν_m conductivity σ, and mean free path l.

Gas	$\mu_e p$, $10^6 \frac{Torr.cm^2}{Vs}$	$\nu_m p$, $10^9 s^{-1} Torr^{-1}$	$\sigma p/n_e$, $10^{-13} \frac{Torr.cm^2}{\Omega}$	range of E/p, $\frac{V}{cm.Torr}$	lp, $10^{-2} cm.Torr$
He	0.86	2.0	1.4	0.6-10	6
Ne	1.50	1.2	2.4	0.2-2	12

For He electron density n_e can be also evaluated as in [15]:

$$n_{e2} = 4.65 \text{x} 10^{13} \frac{j.p_{He}}{E_z} \tag{2}$$

where n_{e2} is in cm^{-3}, j is in A.cm^{-2}, p_{He} is in kPa and E_z is in V.cm^{-1}.

Pulses of the tube voltage U, the discharge current I_R, and the current I_{C3} of the capacitor C_3 are shown in figure 2 (a) for a nanosecond pulsed longitudinal discharge in He used for excitation of $HeHg^+$ laser. Waveforms of the active voltage, the discharge current, the plasma resistance, and the electric power are also presented in figure 2 (b). Time dependence of the electron temperature T_e, the longitudinal electric field E_z, and the electron density n_e is plotted for this gas discharge in figure 3. The electron density is calculated by the two approaches mentioned above. The discrepancy is insignificant.

Waveforms of the active voltage U_R, the discharge current I_R, the plasma resistance R, and the electric power W are given in figure 4 for a nanosecond pulsed longitudinal discharge in He, used for excitation of $He\text{-}Sr^+$ recombination laser.

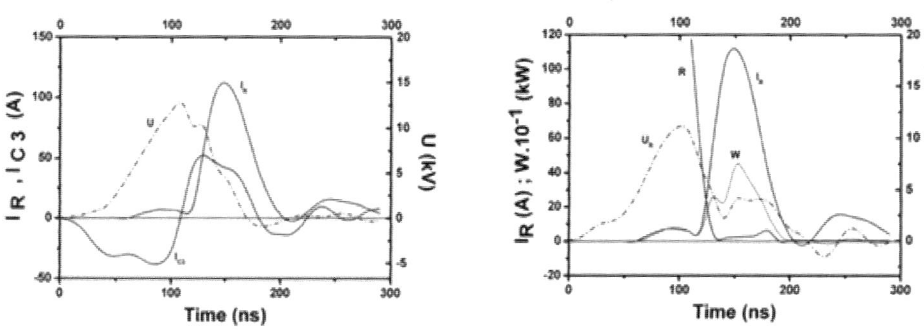

(a) (b)

Figure 2 (a) and (b). Pulses of U, I_R, and I_{C3}, as well as waveforms of U_R, I_R, R, and W for gas discharge in He, exciting $He\text{-}Hg^+$ laser

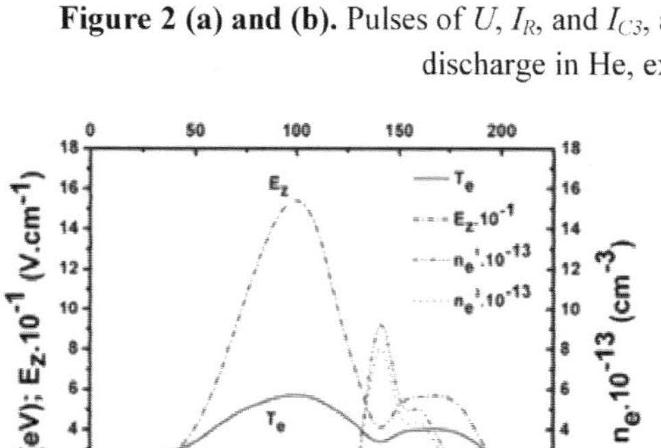

Figure 3..T_e E_z and n_e for gas discharge in He, exciting $He\text{-}Hg^+$ laser

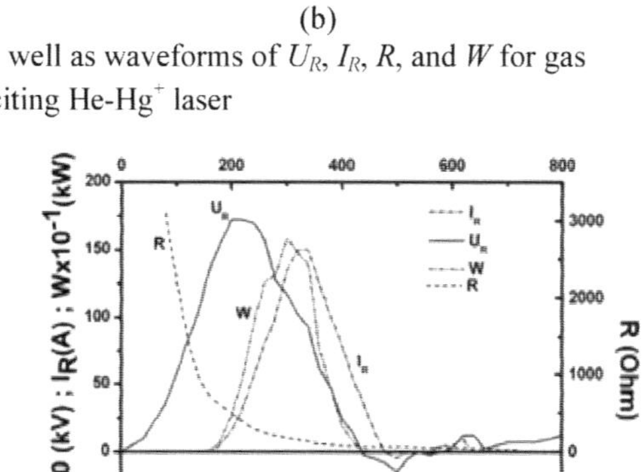

Figure 4. Waveforms of U_R, I_R, R and W for gas discharge in He, exciting $He\text{-}Sr^+$ laser

Pulses of the active voltage U_A and the discharge current I_R (a), as well as waveforms of the plasma resistance R and the electric power W (b), are presented in figure 5 for gas discharge in Ne used for excitation of DUV Cu^+ Ne-CuBr laser.

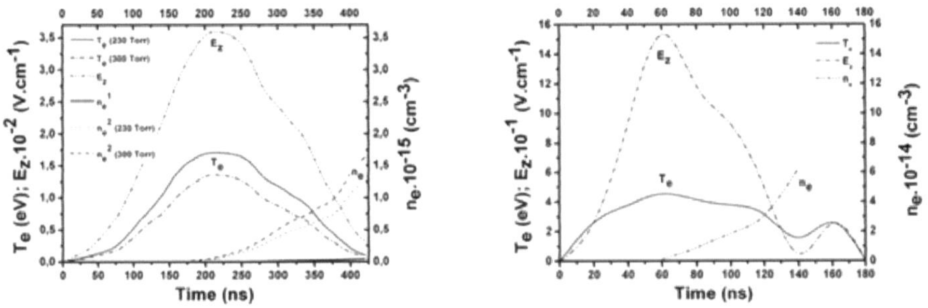

(a) (b)

Figure 5(a) and (b). Waveforms of active voltage U_A and I_R, as well as of R, and W for gas discharge in Ne, exciting DUV Cu^+ Ne-CuBr laser.

Time dependence of the electron temperature T_e, the longitudinal electric field E_z, and the electron density n_e is shown in figure 6 for nanosecond pulsed longitudinal discharge in He (a) and Ne (b) used for excitation of Sr+ recombination and DUV Cu+ Ne-CuBr lasers, respectively

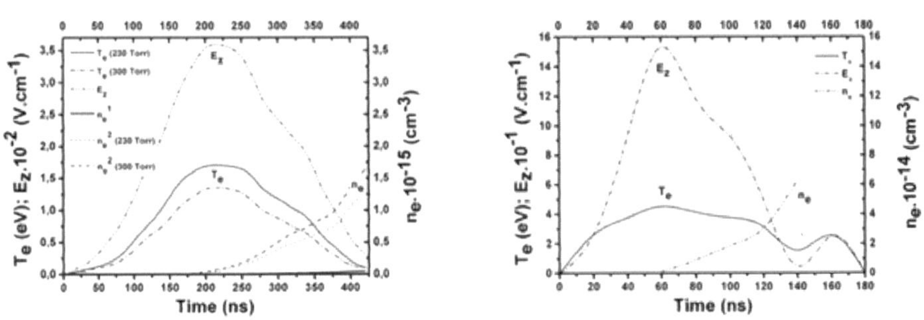

(a) (b)

Figure 6 (a) and (b). T_e, E_z, and ne for gas discharge in He and Ne, respectively.

4. Conclusions

The electrical properties of NPLDs, exciting the high-power DUV Cu^+ Ne-CuBr, $He\text{-}Hg^+$, and $He\text{-}Sr^+$ lasers, are investigated. The time-resolved measurement of electrical discharge characteristics, such as tube voltage and discharge current, permits determination of electron temperature and electron density in the discharge period of these gas discharges. The obtained results are in fairly good agreement with the existing self-consistent models for the DUV Cu^+ Ne-CuBr and $He\text{-}Sr^+$ lasers [1-3, 18] or with the theoretical estimations, which are made following [17].

Journal of Physics: Conference Series **715** (2016) 012004 doi:10.1088/1742-6596/715/1/012004

References

[1] Carman R J 1990 *IEEE J. Quantum Electronics* **26(9)** 1588-608

[2] Carman R J, Brown J W and Piper J A 1994 *IEEE J. Quantum Electronics* **30(8)** 1876-94

[3] Vuchkov N K, Astadjov D N and Sabotinov N V 1994 *IEEE J. Quantum Electronics* **30(3)** 750-8 [4] Temelkov K A, 1995 *MS thesis* Sofia University

[4] Vuchkov N K, Sabotinov N V and Temelkov K A 1996 *Proc. of SPIE* vol 3052, ed P A Atanasov et al (Bellingham: SPIE) pp 136-9

[5] Vuchkov N K, Temelkov K A and Sabotinov N V 1998 Optics and Laser Technology **30(6-7)** 443- 5

[6] Temelkov K A, 2000 *PhD thesis* Sofia, Bulgaria

[7] Vuchkov N K, Temelkov K A and Sabotinov N V 1999 *IEEE J. Quantum Electronics* **35(12)**

[8] 1799-804

[9] Vuchkov N K, Temelkov K A Zahariev P V and Sabotinov N V 2001 *IEEE J. Quantum Electronics* **37(12)** 1538-46

[10] Temelkov K A, Vuchkov N K, Freijo-Martin I and Ekov R P 2010 *J. Phys. D: Appl. Phys* **43**

[11] 075206

[12] Temelkov K A, Slaveeva S I and Vuchkov N K 2011 *IEEE Transactions on Plasma Science*

[13] **39(3)** 831-5

[14] Slaveeva S I, Temelkov K A and Vuchkov N K 2012 *Journal of physics: Conference Series* **356**

[15] 012011

[16] Vuchkov N K, Temelkov K A and Sabotinov N V 2005 *Proc. of SPIE* vol 5830, ed P A Atanasov et al (Bellingham: SPIE) pp 531-5

[17] Astadjov D N, Vuchkov N K and Sabotinov N V 1988 *IEEE J. Quantum Electronics* **24(9)** 1927-

[18] 35

[19] Ivanov I G, Latush E L and Sem M F 1996 *Metal Vapour Ion Lasers* (Chichester: John Wiley &

[20] Sons)

[21] Granovskii V A 1971 *Electric Current in Gases. Breakdown*, (Moscow: Nauka) (in Russian)

[22] Raiser Yu P 1991 *Gas Discharge Physics*, (Berlin: Springer-Verlag)

[23] Stoilov V, 1995 *MS thesis* Sofia University

Journal of Physics: Conference Series **715** (2016) 012005 doi:10.1088/1742-6596/715/1/012005

2D numerical modelling of gas temperature in a nanosecond pulsed longitudinal He-$SrBr_2$ discharge excited in a high temperature gas-discharge tube for the high-power strontium laser

T P Chernogorova[1], K A Temelkov[2], N K Koleva[2] and N K Vuchkov[2]

[1]Faculty of Mathematics and Informatics, Sofia University, 5 James Bourchier Blvd., 1164 Sofia, Bulgaria

[2]Metal Vapour Lasers Laboratory, Institute of Solid State Physics, Bulgarian Academy of Sciences, 72 Tzarigradsko Chaussee, 1784 Sofia, Bulgaria

E-mail: temelkov@issp.bas.bg

Abstract. An active volume scaling in bore and length of a Sr atom laser excited in a nanosecond pulse longitudinal He-$SrBr_2$ discharge is carried out. Considering axial symmetry and uniform power input, a 2D model (r, z) is developed by numerical methods for determination of gas temperature in a new large-volume high-temperature discharge tube with additional incompact ZrO_2 insulation in the discharge free zone, in order to find out the optimal thermal mode for achievement of maximal output laser parameters. A 2D model (r, z) of gas temperature is developed by numerical methods for axial symmetry and uniform power input. The model determines gas temperature of nanosecond pulsed longitudinal discharge in helium with small additives of strontium and bromine.

1. Introduction

It is well known that gas temperature is one of the basic gas-discharge plasma parameters, which determines heavy particle interaction, heavy particle concentration, gas discharge stability, etc. Therefore experimental and theoretical determination of gas temperature is obviously significant and fundamental and is applied to problems in various fields, such as gas-discharge laser physics, gaseous discharges, plasma technologies, gas-discharge mass spectroscopy, absorption and emission spectroscopy and plasma in general. It is also well known that techniques widely used for gas temperature measurement based on measurements of spectral lines Doppler broadening and thermal lens focal distance are definitely imprecise, i. e. with inadmissible experimental error.

Assuming that the gas temperature varies only in the radial direction and considering uniform and non-uniform power input, gas temperature distributions are calculated through analytical solution of steady state heat conduction equation for nanosecond pulsed longitudinal discharge in series of gas discharge tube constructions [1, 2]. Unfortunately, despite the tremendous efforts, analytical solution of the abovementioned steady-state heat conduction equation for 2D (r, z) and 3D (r, ϕ, z) cases encountered some insuperable obstacles.

2. Experimental setup

Schematic diagram of a new high-temperature large-volume discharge tube is shown in figure 1. The basic tube with a 71.5-mm inside diameter and a 76-mm outside diameter is made of fused quartz. A ceramic insert, confining the discharge zone, with a 30.5-mm inside diameter and a 38.5-mm outside diameter and a length of 98 cm is coaxially sleeved in the basic tube. The zone between the ceramic insert, basic quartz tube and the holders of the ceramic insert is incompactly filled with ZrO_2 fibrous insulation. The discharge operates in a self-heating regime. The temperature at the quartz tube surface is measured by a thermocouple.

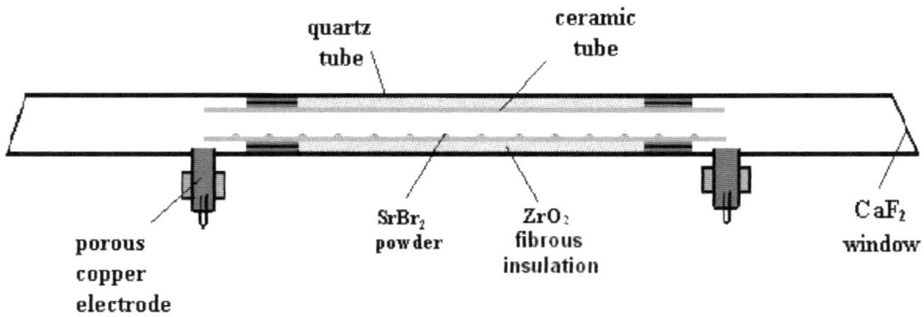

Figure 1. Discharge tube construction.

The electrodes of the discharge tubes are made of porous copper with a special design. CaF_2 windows are stuck to the end of the discharge tubes. The investigated nanosecond pulsed longitudinal discharges are excited by an electrical scheme based on High-Voltage rectifier and High-Voltage pulsed excitation scheme with interacting circuits (IC scheme).

3. Results and discussion

2D numerical solution of the steady-state heat conduction equation is found out for the case of one discharge zone with radius R_1, and three discharge-free zones, namely ceramic tube within $R_1 \le r \le R_2$, discharge-free zone within $R_2 \le r \le R_3$ incompactly filled with ZrO_2 fibrous insulation, and basic tube made of quartz within $R_3 \le r \le R_4$. The dependence of the thermal conductivity k has the form $k = B.T_g^a$, where B and a are constants (within a certain temperature range), which are specific for each gaseous or solid medium. The constants B and a, which determine the thermal conductivity, could be obtained through fitting the existing experimental data taken from [3]. The required thermal conductivities are presented in table 1.

Table 1. $k = B.T_g^a$ – thermal conductivity in $W.m^{-1}.K^{-1}$..

	He	Al_2O_3	ZrO_2	He-ZrO_2	quartz
B	34.9x10^{-4}	44323.1	7326.2x10^{-4}	655.9x10^{-4}	705.9x10^{-4}
a	0.670	-1.227	0.130	0.366	0.487

For the considered four zones 2D heat conduction equations have the following form:

$$\frac{1}{r}\frac{\partial}{\partial r}\left(rk_i(T^{(i)})\frac{\partial T^{(i)}}{\partial r}\right)+\frac{\partial}{\partial z}\left(rk_i(T^{(i)})\frac{\partial T^{(i)}}{\partial z}\right)+q_v=0,\ \ R_{i-1}<r<R_i,\ \ 0<z<L,\ \ R_0=0, \quad (1)$$

where $q_v = 1.9$ $W.cm^{-3}$ is the power deposited into the discharge zone per unit volume and i = 1, 2, 3,

The equations are solved, considering the following boundary conditions:

$$\left.\frac{\partial T^{(1)}}{\partial r}\right|_{r=0} = 0, \quad \left.\mathrm{T}^{(4)}\right|_{r=0} = T_w, \quad 0 < z < L$$

and for i=1,2,3

$$k_i(T^{(i)})\left.\frac{\partial T^{(i)}}{\partial r}\right|_{R_i-0} = k_{i+1}(T^{(i+1)})\left.\frac{\partial T^{(i)}}{\partial r}\right|_{R_i+0}, \quad 0 < z < L$$

$$\left.T^{(i)}\right|_{z=0} = \left.T^{(i)}\right|_{z=L} = T_{w,i}, \; R_{i-1} \le r < R_i,$$

$$\left.T^{(i)}\right|_{z=0} = \left.T^{(i)}\right|_{z=L} = 0.5(T_{w,i} + Tw,i+1), \; r = R_i,$$

$$\left.T^{(4)}\right|_{z=0} = \left.T^{(4)}\right|_{z=L} = T_{w,4}, \; R_3 \le r < R_4$$

The formulated problem is solved, using homogeneous difference schemes and iteration procedure for solving the obtained non-linear system of algebraic equation.

Gas discharge distribution over axial (a) and radial (b) distances are shown for nanosecond pulsed longitudinal discharges in He (45 Torr) in figure 2 and in He (45 Torr) with small admixtures of strontium (0.6 Torr) and bromine (1.2 Torr) in figure 3 and 4, respectively.

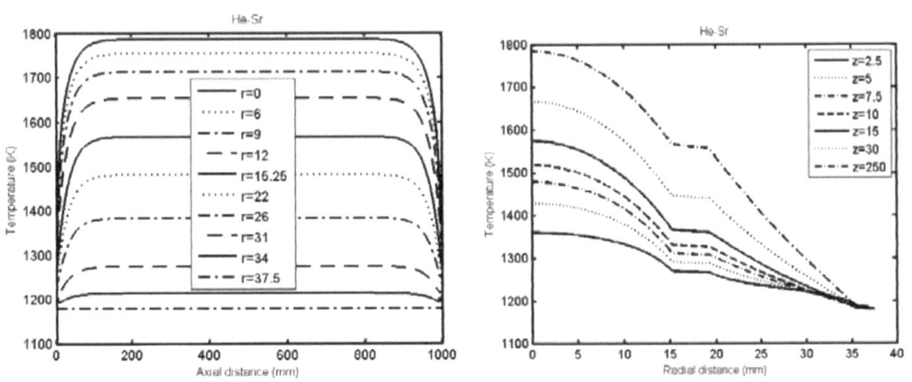

(a) (b)

Figure 2 (a) and (b). Gas discharge distribution over axial (a) and radial (b) distances for nanosecond pulsed longitudinal discharges in He (45 Torr).

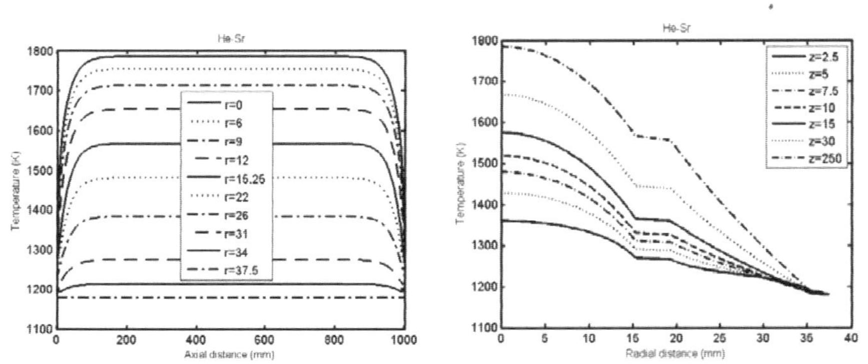

(a) (b)

Figure 3 (a) and (b). Gas discharge distribution over axial (a) and radial (b) distances for nanosecond pulsed longitudinal discharges in Sr-He (0.6 Torr - 45 Torr) mixture.

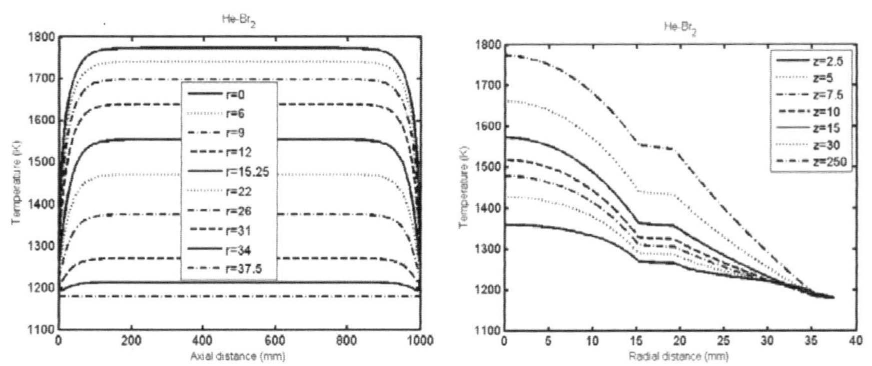

(a) (b)

Figure 4 (a) and (b). Gas discharge distribution over axial (a) and radial (b) distances for nanosecond pulsed longitudinal discharges in Br-He (1.2 Torr - 45 Torr) mixture.

4. Conclusions

A new discharge tube construction with incompact ZrO_2 active-volume insulation of the discharge free zone is developed, in order to increase additionally the operation temperature, and its experimental investigation is in progress. Considering axial symmetry and uniform power input, results from 2D numerical modeling (r, z) of gas temperature in nanosecond pulsed longitudinal discharge in various mixtures are presented for the large-volume high-temperature gas discharge tube newly designed. Development of 2D numerical model for determination of gas temperature in this gas discharge is further more complicated step in approach the reality of experimental conditions for the discharge studied.

Acknowledgements

The first author is supported by the Bulgarian Science Fund for Sofia University under Grant No 181/2012.

References

[1] Temelkov K A, Vuchkov N K, Ekov R P and Sabotinov N V 2008 *J. Phys. D: Appl. Phys* **41** 105203
[2] Temelkov K A, Vuchkov N K, Freijo-Martin I and Ekov R P 2010 *J. Phys. D: Appl. Phys* **43** 075206
[3] Handbook 1976 *Tables of physical constants*, ed I K Kikoin (Moscow: Atomizdat) pp 257-61

Effect of Hydrogen Plasma on Model Corrosion Layers of Bronze

P Fojtíková[1], V Sázavská[1], F Mika[2], F Krčma[1]

[1]Faculty of Chemistry, Brno University of Technology, Purkyňova 118, Brno 612 00, Czech Republic

[2]Institute of Scientific Instruments of the Academy of Sciences Czech Republic, v.v.i., Královopolská 147, Brno 612 00, Czech Republic

E-mail: xcfojtikovap@fch.vutbr.cz

Abstract. Our work is about plasmachemical reduction of model corrosion layers. The model corrosion layers were produced on bronze samples with size of 10 x 10 x 5 mm^3, containing Cu and Sn. Concentrated hydrochloric acid was used as a corrosive environment. The application of reduction process in low-pressure low-temperature hydrogen plasma followed. A quartz cylindrical reactor with two outer copper electrodes was used. Plasma discharge was generated in pure hydrogen by a RF generator. Each corroded sample was treated in different conditions (supplied power and a continual or pulsed regime with a variable duty cycle mode). Process monitoring was ensured by optical emission spectroscopy. After treatment, samples were analyzed by SEM and EDX.

1. Introduction

Beginning of plasma treatment of various subjects is dated on eighties of the twentieth century when group around Daniels used low-pressure hydrogen plasma for the promising treatment of Daguerreotypes and tarnished silver objects [1]. The first success for iron artifacts was achieved by prof. Vepřek around year 1980. He used RF excited hydrogen plasma. Our experimental equipment was constructed according to the design of prof. Vepřek [2], and was further improved.

Together with mechanical cleaning, desalination and surface treatment, hydrogen plasma treatment is a part of conservation process leading to protection of archeological artifacts. Advantages of the method are non-destructivity and no contact with aggressive chemicals. The conservation process has to be performed very specifically because of uniqueness of every archeological artifact. It is not possible to perform research on real artifacts therefore the model samples are necessary. Nowadays, hydrogen plasma treatment is commonly used for iron artifacts. For copper alloys, especially bronze, the optimized conditions during the treatment have not been determined yet. The aim of this study is to determine save conditions of plasma treatment on bronze model samples.

2. Experiment

2.1. Preparation of model corrosion layers

Bronze blocs with the size of 10 x 10 x 5 mm^3 were grinded by an electric grinder because of the unification of roughness of all samples. Sandpaper with roughness of 600 grains per inch square was used. Immediately, the exposure of the corrosion environment followed. Hydrochloric acid was chosen as a model corrosive environment because chlorine compounds play a typical role as bronze corrosion accelerators.
Grinded samples were exposed to vapor of hydrochloric acid in a dessicator. The Petri dish was placed to the bottom of the dessicator, and 20 ml of concentrated hydrochloric acid were infused into it. The perforated ceramic holder with samples was above the Petri dish. Closed dessicator was placed in a dark place, and the development of corrosion was observed. After 14 days, surfaces of all samples were covered with corrosion products, and the corrosion process was stopped.

2.2. *The experimental setup*

The experimental setup is shown in the Figure 1. The quartz cylindrical reactor (length of 90 cm, inner diameter of 9.5 cm) is a base of the plasmachemical apparatus. Other parts are connected to the reactor: vacuum system, source of working gas, source of energy, system of monitoring process (optical emission spectroscopy) and a pressure gauge. The vacuum system is used to product vacuum of about 10 Pa, and it is composed of a rotary oil pump, an electromagnetic and a ball valve. Pure hydrogen was chosen as working gas, and it was fed into the reactor through a mass flow controller. The RF generator supplies energy needed for the plasma discharge creation (it works on the technical frequency of 13.56 MHz). The generator is connected to two grounding copper electrodes placed out of the cylindrical reactor by a fiberglass. Plasmachemical process of corrosion layers removal is monitored by optical emission spectroscopy. Signal from the reactor is routed through the quartz slot to the optical fiber. The optical fiber is connected with a spectrometer Ocean Optics HR 4000. The experiment is PC controlled.

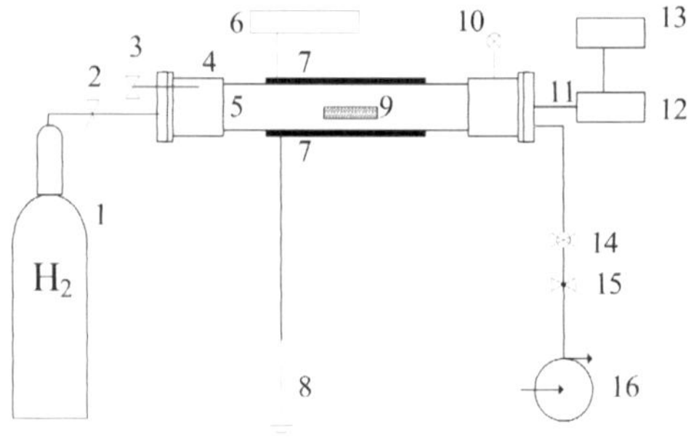

Figure 4. Experimental setup: 1 – hydrogen, 2 – mass flow controller, 3 – aeration valve, 4 – reactor, 5 – holder, 6 – RF generator, 7 – 2 copper electrodes, 8 – grounding, 9 – sample, 10 – pressure gauge, 11 – optical fiber, 12 – spectrometer, 13 – computer, 14 – ball valve, 15 – electromagnetic valve, 16 – rotary oil pump.

2.3. Analysis

The plasma treated samples were analyzed by a scanning electron microscope (SEM) Magellan 400L in order to obtain ultra high resolution surface morphology and chemical composition by energy dispersive x-ray analysis (EDX). SEM pictures were taken at the energy of 5 keV and 15 keV with the magnification of 1000–1200. EDX analysis was performed from the area of 5x5 μm^2.

2.4. The course of experiment

For each sample, different conditions of supplied power (from 50 to 400 W) and the duty cycle mode (Table 1) were chosen. During the experiment, the pressure in the reactor ranged about 150–170 Pa

and the flow rate of hydrogen was 50 sccm. The corroded sample was put on the glass holder in the middle of the reactor between two electrodes.
The active time of discharge was adjustable in the pulsed mode, expressed as a percentage

$$DUTY\ CYCLE = \frac{t_{on}}{t_{on} + t_{off}} \cdot 100\ [\%], \quad (1)$$

where t_{on} / t_{off} is time when the discharge was switched on/off [ms]. Time of 1 pulse was 1 ms because the frequency of repetition of 1 pulse was chosen as 1000 Hz. Because of pulse mode, the power (supplied by RF generator) is reduced to effective power which is calculating using the equation

$$P_{eff} = P \cdot DUTY\ CYCLE\ [W], \quad (2)$$

In continual mode, the power is the same as an effective power because the discharge is switched on all time [3].

Table 1: Conditions during plasmachemical reduction for all 10 samples.

SAMPLE	POWER [W]	MODE*	DUTY CYCLE [%]	EFFECTIVE POWER [W]
1	50	C	100	50 (Group A)
2	100	P	50	
3	200	P	25	
4	100	C	100	100 (Group B)
5	200	P	50	
6	300	P	33	
7	400	P	25	
8	200	C	100	200 (Group C)
9	300	P	66	
10	400	P	50	

*Note: C – continual, P – pulsed.

3. Results and discussion

3.1. Time dependence of OH radical emission

Atomic hydrogen is one of the products created by plasma. If atomic hydrogen reacts with oxygen in the corrosion layer, OH radicals are produced. This running process was monitored by the optical emission spectroscopy.

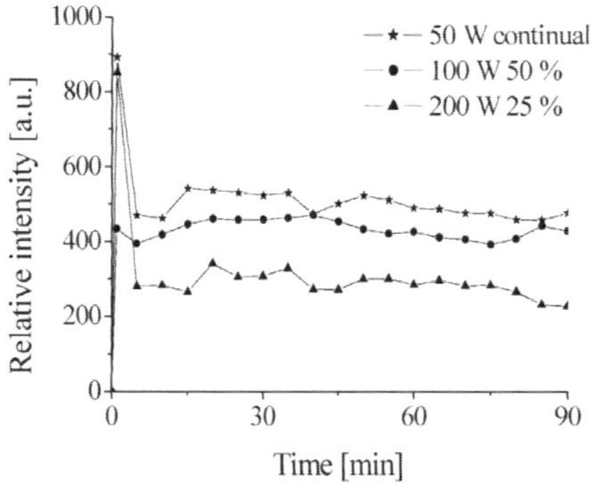

Figure 2. Time dependence of relative intensity of OH radicals for Group A samples.

Figure 2 shows the dependences of the relative intensity of OH radicals on time at different conditions. We can see an increase of the intensity in the first 5 minutes. When the intensity maximum is reached, the decrease followed. The process was stopped after 90 minutes because of a steady tendency of the relative intensity of OH radicals.

3.2. SEM and EDX analyses

All ten samples were analyzed by SEM. For illustration, sample 8 was chosen (Group C) and the picture of its surface as well as the result of the EDX analysis is shown below.

Figure 3. Corroded bronze sample 8 before (left) and after (right) the treatment, recorded by SEM. The red square is an area 5x5 μm^2 from which the EDX analysis was done (Figure 4, Table 2).

From the EDX results (Table 2), we can see that the mass of oxygen and chlorine is decreasing. This effect is desirable because due to the decrease of oxygen (formation of OH radicals by the reaction with atomic hydrogen), it is possible to monitor the experiment. Reduced mass of chlorine is also a benefit because chlorine is the main agent of the soil corrosion.

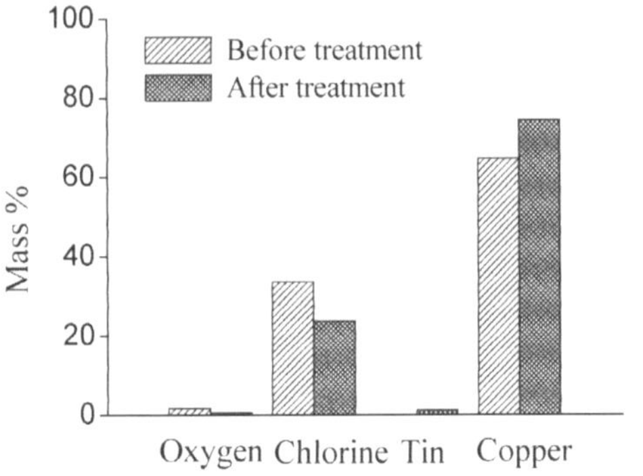

Figure 4. Mass % of the sample 8 (EDX).

	Before treatment	
	Mass %	**Atomic %**
Oxygen	1.67	5.04
Chlorine	33.55	45.71
Tin	- - -	- - -
Copper	64.78	49.25
	After treatment	
	Mass %	**Atomic %**
Oxygen	0.56	1.87
Chlorine	23.76	35.5
Tin	1.22	0.54
Copper	74.46	62.09

Table 2. Results of EDX analysis of sample 8.

4. Conclusion

The plasmachemical reduction of bronze samples with model corrosion layers was performed. The results of OES show that the process of reduction began in the first two minutes of treatment. From EDX analysis results, it is evident that there was a decrease of oxygen and chlorine. Decrease of chlorine is eligible because chloride ions are the main agents of the soil corrosion. This method is a part of an archeological approach at conservation of artifacts. The main advantage is its non-destructivity and no contact with aggressive chemicals.

Acknowledgements

This research has been conducted within the project DF11P01OVV004 „Plasma chemical processes and technologies for conservation of archaeological metallic objects“, founded by the Ministry of Culture of the Czech Republic.

References

[1] Daniels V. D.; Holland L.; Pascoe M. W.: Gas plasma reactions for the conservation of antiquities. *Studies in Conservation* 1979, 24, 85–92.

[2] Vepřek, S.; Patscheider, J.; Elmer, J.: Restoration and conservation of Ancient artifacts: A new area of application of plasma chemisty. *Plasma Chemistry and Plasma Processing* 1985, 5, 201–209.

[3] Zemánek, N.: Study of plasmachemical reduction of corrosive layers on. *Diploma thesis* 2008, VUT Brno.

Journal of Physics: Conference Series **715** (2016) 012007 doi:10.1088/1742-6596/715/1/012007

Discharge ignition in the diaphragm configuration supplied by DC non-pulsing voltage

L. Hlochová, L. Hlavatá, Z. Kozáková, F. Krčma

Faculty of Chemistry, Brno University of Technology, Purkyňova 118, Brno 612 00, Czech Republic

E-mail: xchlochova@fch.vutbr.cz

Abstract. This work deals with the ignition of the discharge in the diaphragm configuration generated in water solutions containing supporting NaCl electrolyte. The reactor has volume of 110 ml and it is made of polycarbonate. HV electrodes made of stainless steel are placed in this reactor. Ceramic (Shapal-M^{TM}) diaphragm is placed in the barrier separating the cathode and the anode space. An electric power source supplies the reactor by constant DC voltage up to 4 kV and electric current up to 300 mA. The discharge ignition is compared in the reactor with different sizes of diaphragms. Measurements are carried out in electrolyte solutions with the same conductivity. Images of plasma streamers and bubble formation are taken by an ICCD camera iStar 734. Electrical characteristics are measured by an oscilloscope LeCroy LT 374 L in order to determine breakdown moments at different experimental conditions.

1. Introduction

Electrical discharges generated directly in liquids or direct plasma interaction with liquid phase is one of the hot topics of the contemporary plasma research. These systems can be applied for example in water treatment [1, 2], sterilization [3], and surface treatment of various temperature sensitive materials [4] as well as for the nanoparticles generation [5]. The plasma in water solutions can be generated by various electrode configurations, and it can be supplied by many kinds of the applied voltage from DC up to RF frequencies using pulsed as well as non-pulsed regimes. In general, there are two theories describing the discharge ignition in liquids.

The electron theory is based on the fact that water molecules are ionized and dissociated by the applied strong electric field, and the plasma creation is very similar to the classical Townsend's theory of electron avalanches in gases [6].

The thermal (bubble) theory supposes that liquid is strongly heated by passing current and thus it is evaporated and micro bubbles are created. The discharge is consequently ignited due to the strong potential gradient inside these bubbles [6].

The presented contribution deals with the direct observation of the breakdown of diaphragm and capillary discharges in water solutions. In this case, both electrodes are immersed inside the liquid phase, and the electrodes are separated by a dielectric barrier with a small orifice. In the case when the aspect ratio (thickness/diameter) is about 1 (thin barrier with a relatively big orifice), the discharge is called the diaphragm discharge; if the aspect ratio is much higher than 1, the discharge is called the capillary discharge. The non-pulsing DC voltage was used for the discharge generation. The time resolved current-voltage characteristics and ICCD imaging were applied for the discharge breakdown determination.

2. Experiment

One batch discharge reactor was used for the presented study. Its volume was about 110 ml of water solution. Its simplified scheme is shown in figure 1.

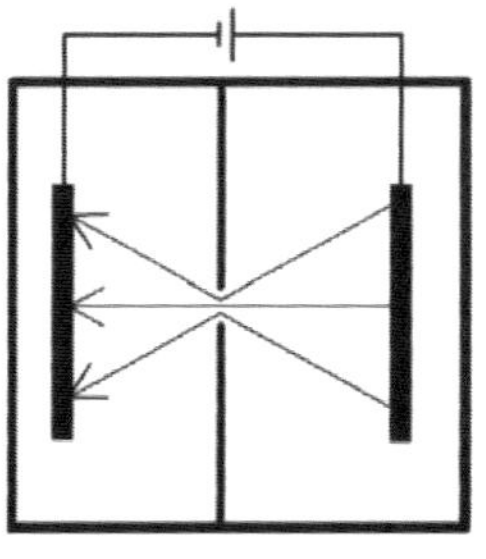

Figure 1. Simplified scheme of the diaphragm/capillary discharge configuration. Arrows represent the electric intensity vector.

The electrodes were made of stainless steel; the diaphragm was made of Shapal-M^{TM} ceramics. A set of diaphragms with one central orifice with the diameter of 0.3 and 0.6 mm and thickness of 0.3–1.5 mm was used for the contemporary experiments. NaCl solutions with conductivity of 250 μS were prepared.

Diagnostics of the system were performed by a four channels LeCroy LT374L oscilloscope and the ICCD camera (Andor iStar with Cosmicar/Pentax TV Lens 50 mm, 1:2.8 objective). The ICCD camera was synchronized with the discharge by a current value.

3. Results

The current-voltage characteristics of the diaphragm discharge breakdown were measured for the ceramic diaphragm (orifice diameter of 0.3 and 0.6 mm) with the thickness of 0.3–1.5 mm. NaCl solutions of 250 μS were used. The current-voltage characteristics for diaphragm discharge with barrier with orifice diameter 0.3 mm and different thickness are shown in figure 2 – left, and with the barrier with orifice diameter 0.6 mm and different thickness are shown in figure 2 – right.

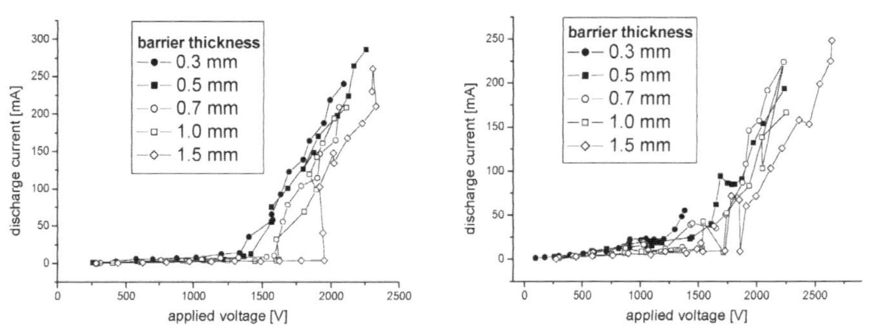

Figure 2. Current-voltage characteristic of the diaphragm discharge breakdown in NaCl solution (250 μS/cm). Ceramic diaphragm (orifice of 0.3 mm – left, orifice of 0.6 mm – right), thickness of 0.3–1.5 mm.

On these current-voltage characteristics it can be observed that in the first part (up to about 700 V), very low current is determined which corresponds to regular electrolysis. Due to the high current density in the orifice, the solution is overheated in the diaphragm, and microbubbles are generated. Resistance between the electrodes increases and thus current passing through the system decreases [7].

When the applied voltage reaches values of about 1200 V, the discharge is ignited inside small bubbles. The discharge lifetime is very short because the bubble rapidly expands, and applied voltage is not sufficient for the discharge maintenance. If the applied voltage is around 1400 V, the discharge becomes more regular, and the current rapidly increases.

Comparing figure 2 left and right, it can be visible that in the case of the bigger orifice (Fig. 2 – right), discharge current reaches lower values with smaller voltage. Increasing the orifice diameter, the shift of curves to higher values of voltage can be observed. Discharges created in the barriers with the bigger thickness require higher voltage at the same current.

In figures 3–6 discharge ignition and the camera closing in the dependence on current is demonstrated. When discharge current increases it is a signal for the camera closing because the discharge is ignited. For smaller diaphragm orifices, higher current is needed for the ignition. On photographs assigned to the graphs, discharges under the barriers are visible. This light emission corresponds to the grey curves in the graphs which are demonstrating the camera closing.

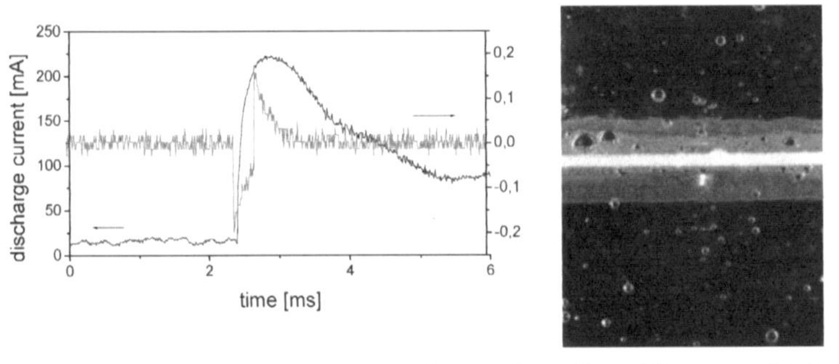

Figure 3. Experiment with the diaphragm with the smaller orifice (diameter of 0.3 mm) and thickness of 0.3 mm. Camera closing by a current impulse (left). Photography exposure time of 0.5 ms (right).

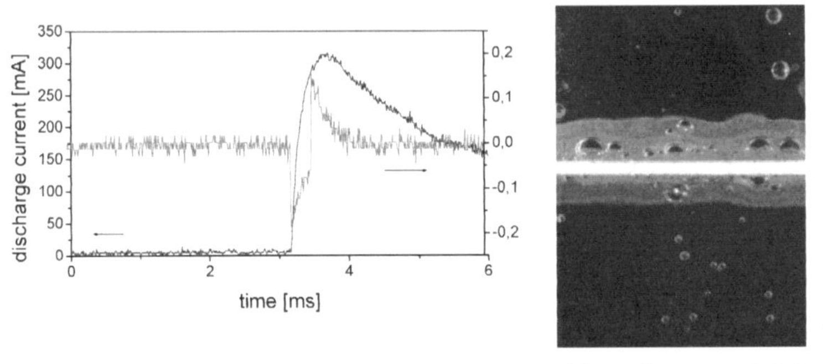

Figure 4. Experiment with the diaphragm with the smaller orifice (diameter of 0.3 mm) and thickness of 1.0 mm. Camera closing by a current impulse (left). Photography exposure time of 0.5 ms (right).

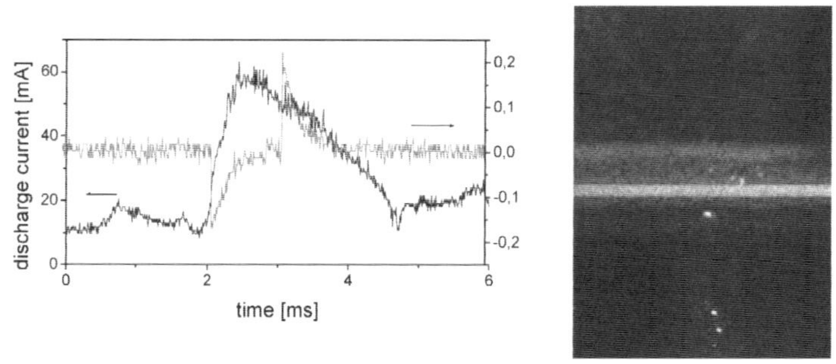

Figure 5. Experiment with the diaphragm with the bigger orifice (diameter of 0.6 mm) and thickness of 0.3 mm. Camera closing by a current impulse (left). Photography exposure time of 1.0 ms (right).

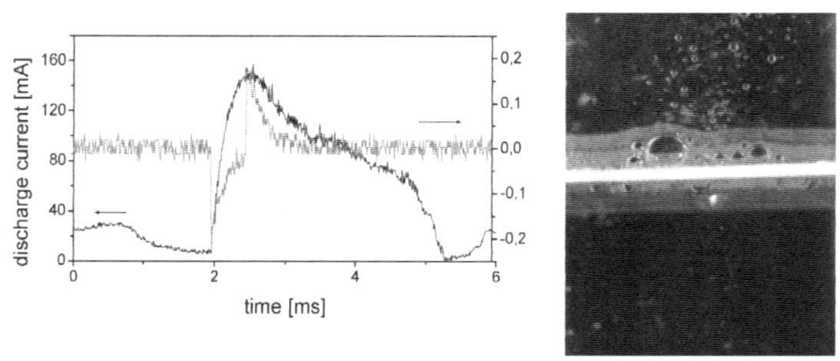

Figure 6. Experiment with the diaphragm with the bigger orifice (diameter of 0.6 mm) and thickness of 1.0 mm. Camera closing by a current impulse (left). Photography exposure time of 0.5 ms (right).

4. Conclusions

The discharge breakdown was observed in the diaphragm and capillary configurations. It was compared in the dependence on the aspect ratio of the diaphragm (thickness and orifice diameter). For smaller orifice diameter, higher current and voltage are needed for the discharge ignition. Increasing the dielectric barrier thickness, discharge current and applied voltage reach higher values.

Acknowledgements

This work has been supported by the Ministry of Culture of the Czech Republic, project No. DF11P01OVV004.

References

[1] Sázavská, V., Kozáková, Z., Hlochová, L., Krčma, F., Slavíček, P., Mazánková, V. ISPIC XX. Philadelphia: IUPAC, 2011. LIQ08
[2] A.T. Sugiarto, S. Ito, T. Ohshima, M. Sato, J.D. Skalný, J. Electrostatics, 58 (2003) 135.
[3] L. Marsili, S. Espie, J.G. Anderson, S.J. MacGregor, Radiat. Phys. Chem., 65 (2002) 507.
[4] M. Klíma, P. Slavíček, M. Šíra, T. Čižmár, P. Vaněk, Czech. J. Phys., 56 (2006) B1051.
[5] S.M. Thagard, K. Takashima, A. Mizuno, Plasma Process. Polym., 6 (2009) 741.
[6] B. Sun, M. Sato, J.S. Clements, J. Electrostatics, 39 (1997) 189.
[7] F. Krčma, L. Hlavatá, L. Hlochová, Z. Kozáková, O. Guaitella, I. Marinov, S. Starikovskaia. ICPIG XXX (2011)

Journal of Physics: Conference Series **715** (2016) 012008 doi:10.1088/1742-6596/715/1/012008

Model of a small surface wave discharge at atmospheric pressure

A Ivanov, Zh Kiss'ovski

Faculty of Physics, Sofia University "St. Kl. Ohridski" 5, J Bourchier Blvd., BG-1164, Sofia, Bulgaria

E-mail: kissov@phys.uni-sofia.bg

Abstract. Self-consistent model of a small microwave plasma source based on a surface wave sustained discharge at 2.45 GHz is presented in this study. The model includes dispersion relation of azimuthally symmetric surface waves, sustaining the discharge in a high permittivity ceramic tube (εd = 9.3) and the radial distribution of the field components at curtain values of the electron density are obtained. The electron Boltzmann equation under the local approximation is solved together with the heavy particle balance equations. A detailed collisional-radiative model for argon discharge at atmospheric pressure is implemented in the model. The changes in the EEDF shape and the mean electron energy with the value of the electron density are investigated. Results show that the EEDF is close to Maxwellian at our experimental conditions for the plasma density above 2.1020 (m-3).

1. Inroduction

The surface wave (sw) discharges at atmospheric pressure are object of intensive investigations due to their applications for environmental and industrial purposes as detoxification of hazardous gases, hydrogen production, plasma chemistry, cutting and surface modification of metals.

Modelling of these discharges is difficult because a large number of reactions between charged particles and neutrals and complexity of the processes even in argon gas [1-3]. Theoretical results for the electron energy distribution function (EEDF) for specific gas discharge conditions [2, 4] show that it can deviate from Maxwellian at low plasma densities in the surface wave discharge at atmospheric pressure. These changes in the shape of EEDF require specification of the range of plasma parameters where the double probe diagnostics and collisional-radiative models which assume Maxellian distribution function can be applied.

This study presents a model of plasma column sustained by sw in a small radius ceramic capillary of portable microwave plasma source [5]. Discharge is self ignited at power levels below 10 W and it works at atmospheric pressure both in continuous and pulse regimes. The model includes dispersion relation of azimuthally symmetric surface waves, sustaining the discharge in a high permittivity ceramic tube (ε_d = 9.3). The electron Boltzmann equation under the local approximation is solved together with the heavy particle balance equations. The optical and double probe measurements of the plasma parameters show that our source produces dense plasmas [6] and detailed collisional-radiative model for argon discharge at atmospheric pressure is implemented in this self-consistent model. The changes in the EEDF shape and the mean electron energy at different values of the electron density are investigated. The electron temperature and populations of 1s2-1s5 and 2p1-2p10 and higher levels of excited argon atoms are determined at our experimental conditions.

2. Description of the plasma source and its model.

The sw exciter is open coaxial line and the dielectric of this line is an alumina ceramic tube with inner diameter of d = 1mm and outer of 2 mm and length of 11 mm. The electric field has two components E_r and E_z at the end of the exciter which cause the breakdown of the gas and propagation of the surface waves. Microwave signal at frequency f = 2.45 GHz from generator MPG-4M (0-120 W) through a double directional coupler Narda-3022 and triple stub Maury 1878C is fed to the source.

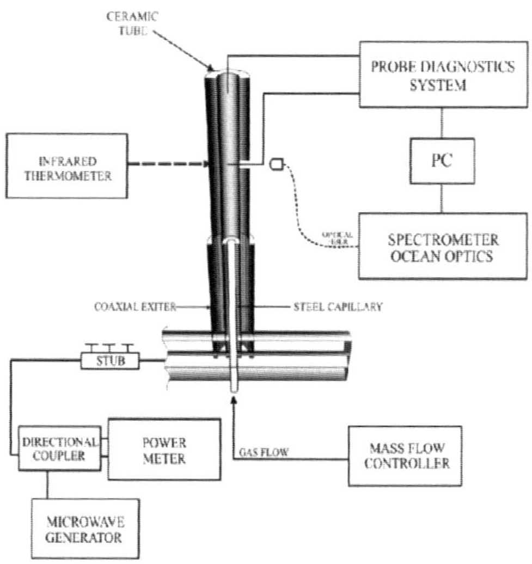

Figure 1. Small surface wave discharge

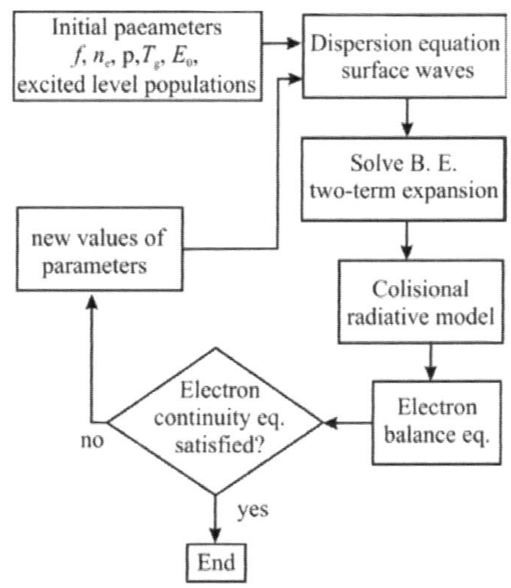

Figure 2. Flow-chart of the model

The discharge is self-ignited at microwave power levels below 10 W and it works in continuous and pulse regimes in argon and argon-helium gas mixture. The length of the plasma column is controlled by the applied microwave power and the source could operate as a plasma jet. Forward and reflected power from the plasma source is measured by power meter HP 437B. Electron temperature (T_e~1.6 eV) and density ($n_e \sim 4.10^{20}$ m^{-3}) are estimated by optical diagnostics and asymmetric double probe characteristics. Optical emission spectroscopy observation of argon discharge are realized by HR Ocean Optics Spectrometer. The light from the plasma column is collected by lens system (through the same small hole) coupled to optical fibre, which is connected to the spectrometer. Spectrometer UV-VIS is used for determination of the neutral gas temperature (T_g~1500 K) from OH-band present in the spectra.

A flow chart of the theoretical model is shown on figure 2. The iterative self-consistent model includes solver for the dispersion equation (DE) for the sw, Boltzmann equation for the electrons, module with collisional-radiative model for the heavy particles and electron particle balance equation. The dispersion equation of sw is solved for the case of thin cylinder approximation in cylindrical coordinates, with waves propagating in axial direction along the plasma column. The components of the azimuthally symmetric mode (TM_{00}) are E_z, E_r H_φ with a maximum of the field amplitude at the plasma-dielectric interface. The plasma column is assumed homogeneous in radial direction and dielectric constant is calculated with averaged value of the electron concentration n_e and effective value of the electron-neutral collision frequency ν_{en}. Applying the boundary conditions across the tree structure domains (plasma-dielectric-air) the dispersion relation is solved numerically via an iterative procedure. The values for the wave damping rate α and the phase constant β are obtained for initial

Journal of Physics: Conference Series **715** (2016) 012008 doi:10.1088/1742-6596/715/1/012008

value of the electron concentration and electron-neutral collision frequency. With the complex wave vector the distribution of the electromagnetic field in radial direction is obtained and also the value of effective electric field at the local position is calculated as $E_0 = \sqrt{E_z^2 + E_r^2}$. The electron Boltzmann equation (BE) under the local approximation is numerically solved for the EEDF using two-term expansion for the stationary regime of the discharge. Initial parameters for this solver are the obtained value for the effective field from the DE module, neutral gas pressure, electron concentration, mean electron energy, gas temperature and initial values of the concentrations of heavy particles.

The following heavy particles are included in the model: Ar s-levels (1s2-1s5), the Ar 2p-levels (2p1 -2p10), blocks for Ar(2s3d) and higher levels, atomic Ar+ and molecular Ar2+ ions and exited molecules Ar2*. The kinetic processes considered in the BE are elastic collisions and inelastic collisions (excitation, de-excitation, ionization, step-wise ionization, dissociative recombination for Ar2+ and tree body recombination etc.) and Penning ionization for the excited molecules.

3. Results

The Boltzmann equation is solved via iterative finite difference method and EEDF and the value of the mean electron energy are obtained. The shape of EEDF is close to Maxwellian for values of the electron density above 2.10^{20} (m^{-3}) (figure 3a). The tail of EEDF at electron densities lower than 2.10^{20} (m^{-3}) significantly deviates and it is below the black line of Maxwellian distribution especialy in the energy range above 11 eV which is in agreement with results in [2]. The mean electron energy (effective electron temperature) calculated from EEDF decreases with increase of the plasma density (Figure 3b) but it is higher than the electron temperature obtained in sw discharges in a quartz tube with same inner diameter. The radial distribution of electron temperature at n_e= $4.10^{20} m^{-3}$ (EEDF is Maxwellian) is presented in figure 4 and it has weak increase in the direction to the tube wall.

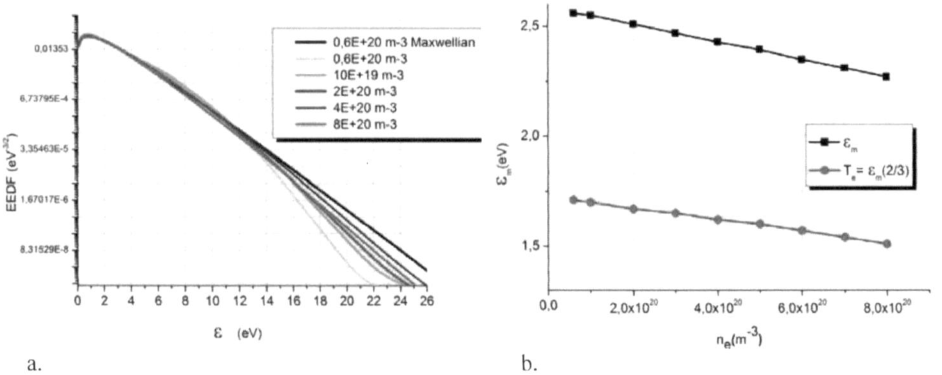

Figure 3. Results for EEDF: (a) shape and (b) mean electron energy at different values of n_e

With obtained EEDF and mean electron energy (electron temperature), electron density and gas temperature we calculate the population of excited levels of argon atoms by collisional-radiative model. Model for argon discharge at atmospheric pressure takes into account the species at s-levels Ar(1s2-1s5), at 2p-levels Ar(2p1 -2p10) and blocks for Ar(2s3d) and higher levels, atomic Ar^+ and molecular Ar_2^+ ions, exited molecules Ar_2* are also implemented in the model. The main kinetic reactions are included in the model: electron-impact excitation from ground state, from 1s levels, the electron-impact de-excitation from 1s, electron-impact population transfer, atom-collision population transfer, electron-impact ionization and Penning ionization of excited particles, electron collisional recombination for atomic and molecular ions, 3-body collisional association reactions, electron-impact and atom-collisional dissociation reactions.

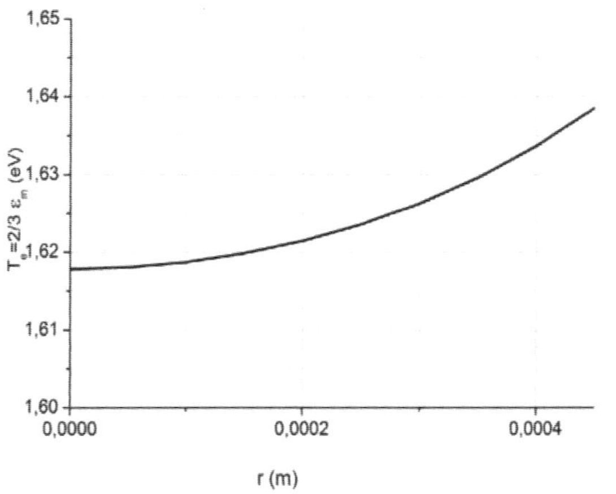

Figure 4. Radial distribution of the electron temperature in the plasma column.

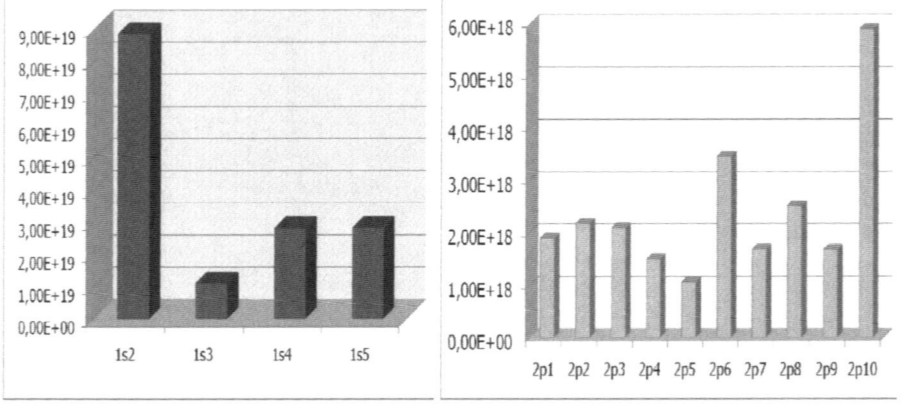

Figure 5. Population densities of : a) 1s b) 2p from CMR at at $n_e = 4.10^{20} m^{-3}$

Results for the absolute population of 1s and 2p levels at our experimental conditions (T_g, T_e, n_e) obtained with this model are presented in figure 4. Populations of 1s and 2p levels of the Ar atoms strongly depend on the plasma density and the mean energy of electrons and their population distribution is far away from equilibrium.

4. Conclusion

The model of surface wave discharge at atmospheric pressure in argon gas in a high permittivity ceramic tube ($\varepsilon_d = 9.3$) is developed. The Boltzmann equition for electrons is solved together with the collisional-radiative model for argon plasma at atmospheric pressure. Result for EEDF shows that it is close to Maxwellian distribution if the plasma density in the capillary is above 2.10^{20} m^{-3}. The mean electron energy and T_e decreases with increase of the plasma density and radial position.

Acknowledgements
This work is financially supported by Science Fund of Sofia University, project N150. The authors thank to Prof. E Tatarova and Dr. J Henriques for their constant interest and support.

References

[1] E. Tatarova, F. M. Dias, E. Felizardo, J. Henriques, M. J. Pinheiro,C. M. Ferreira,B. Gordiets, 2010 *J. of Applied Phys.* **108**, 123305
[2] Pencheva M, Petrov G, Petrova T, Benova E, 2004 *Vacuum. **76**, 409*
[3] Pencheva M, Petrova T, Benova E, Zhelyazkov I, 2008 *J. of Phys. Conf. Ser. **44**, 11*
[4] Henriques J, Ivanov A, Tatarova E, Kiss'ovski Zh, Ferreira C M, *Bulletin APS*, **56**, 35
[5] Kiss'ovski Zh, Kolev M, Ivanov A, Lishev St , Koleva I, 2009 *J Phys. D: Appl. Phys*., **42,** 182004
[6] Kiss'ovski Zh, Ivanov A, Iordanova S, Koleva I, 2011 *J Phys. D: Appl. Phys*., **44,** 205203

Journal of Physics: Conference Series **715** (2016) 012009 doi:10.1088/1742-6596/715/1/012009

A novel device for spectrochemical analysis based on a combination of LIBS and a hollow cathode discharge

S Karatodorov[1], V Steflekova[1], V Mihailov[1], D Mihailova[2], J van Dijk[2], J van der Mullen[2] and M Grozeva[1]

[1]Institute of Solid State Physics, Bulgarian Academy of Sciences, Bulgaria

[2]Department of Applied Physics, Eindhoven University of Technology, The Netherlands

E-mail: skarat@issp.bas.bg

Abstract. A new modified analytical scheme of laser ablation-hollow cathode discharge (LA- HCD) for elemental analysis is reported. The novelty is the use of a hollow cathode with a longitudinal slit. This slit enables registration of the axial species distribution in the discharge and decoupling of the laser and the analytical signals. In this paper preliminary experiments on the modified LA-HCD set-up are reported. The improved analytical performance of the new LA-HCD scheme is demonstrated by comparing it to stand-alone laser induced breakdown spectroscopy. The results of the modified LA-HCD technique show enhancement of the excitation efficiency.

1. Introduction

Spectrochemical analysis is a powerful instrumental method for the determination of the elemental composition of various materials. From the variety of methods for spectrochemical analysis, one that is being widely applied in recent years is laser induced breakdown spectroscopy (LIBS) [1]. The LIBS technique offers the opportunity to analyze any material and to detect all elements in the periodic table, requiring at the same time no or minimal sample preparation. These features enable the technique to perform analysis of multi-component samples in the laboratory or in-situ in a fast automated measurement process. The main drawbacks of the technique are the matrix effects and the lower sensitivity compared to other techniques [2]. One of the possible ways to overcome these drawbacks is to combine the LIBS technique with a gas discharge [3]. In such a combined technique the laser provides ablation and atomization of the sample and the gas discharge provides excitation and/or ionization of the ablated atoms. Such hybrid laser-gas discharge schemes employ a variety of different discharges - ICP, microwave discharge, spark/arc discharge, DC plasma devices, hollow cathode discharge (HCD). One hybrid technique that is being reported in recent years is laser ablation-hollow cathode discharge (LA- HCD) technique [4-7]. In the existing realizations of the technique, the HCD is operated in different electrical regimes - DC, RF and pulse mode and the registration of the discharge emission is done on-axis.

In this work a novel modification of the LA-HCD technique is presented, allowing also observation of the side emission from the discharge. This is made possible by the use of a hollow cathode (HC) with a slit cut along the axis. The capabilities of the modified LA-HCD are demonstrated by comparing its results with those of stand-alone LIBS.

2. Experimental setup and conditions

The experimental setup of the modified LA-HCD technique is shown in figure 1. It consists of three main parts: HCD, laser and signal registration.

The modified HC is a hollow cylinder with a longitudinal slit. The slit enables registration of the discharge side emission. Such registration gives opportunity for obtaining the species axial distribution and the evolution of this distribution in time. Apart from that, this slit allows the decoupling of the laser excitation and the analytical signal acquisition. In this way this novel LA-HCD scheme widens the scope of applications of the LA-HCD technique and makes it easier to perform.

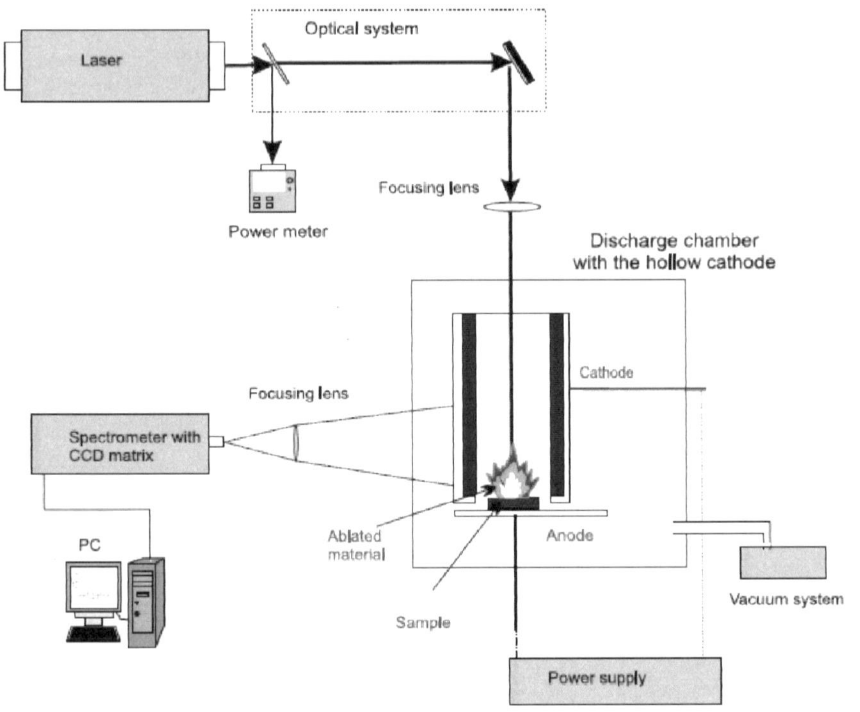

Figure 1. Experimental setup of the LA-HCD technique.

The discharge part of the set-up consists of the discharge chamber, the HC, the power supply and the vacuum system. The dimensions of the HC are inner radius 4 mm and length 20 mm. These dimensions are chosen according to results from PLASIMO's MD2D model for homogeneous electron density distribution [8-10]. The HC is made from duralumin because this material has small sputtering yield. The anode is a disc. The sample is standing on top of the anode or can be the anode itself, if the sample is made from conducting material. HC is in a chamber with controlled pressure and gas composition. For the first experiments helium is chosen as buffer gas, as it causes low sputtering of the cathode wall. The discharge is excited by DC power supply.

The laser source is Quanta Ray GCR3 pulsed Nd:YAG laser. The pulse repetition rate of the laser can be changed from 1 to 30 Hz, the pulse duration for the primary wavelength 1064 nm is 8 ns and the laser pulse energy is between 10 ÷ 450 mJ. The spectroscopic part of the system consists of a Digikrom 240 spectrometer controlled by a PC.

The laser beam is directed to the HC by a set of mirrors and through a lens is focused onto the sample material, which is placed on the anode. The focused laser beam ablates atoms from the surface of the material. The ablated atoms enter the HCD and there they are excited and/or ionized. The optical emission from these atoms is then registered through the longitudinal slit of the HC and focused by the collective lens on the entrance slit of the spectrometer.

The experiments are performed with the fundamental wavelength of the Nd:YAG laser, that is 1064 nm , with laser repetition rate of 10 Hz and with laser pulse energy of 80 mJ. The buffer gas is helium with pressure of 4.2 Torr. The discharge current is varied from 10 to 40 mA.

3. Experimental results

The first experiment performed on this setup is the comparison of the modified LA-HCD scheme to stand-alone LIBS. The investigated sample is the copper anode. The anode is made with radius bigger than that of the HC and it can be rotated, so that the laser beam does not hit each time at the same point on the anode. The spectra obtained under different excitation and conditions are compared.

Figure 2 shows the obtained spectrum of the investigated copper sample. The main figure shows the full spectrum obtained by stand-alone LIBS in helium atmosphere (4.2 Torr) over the wavelength range 300÷800nm.

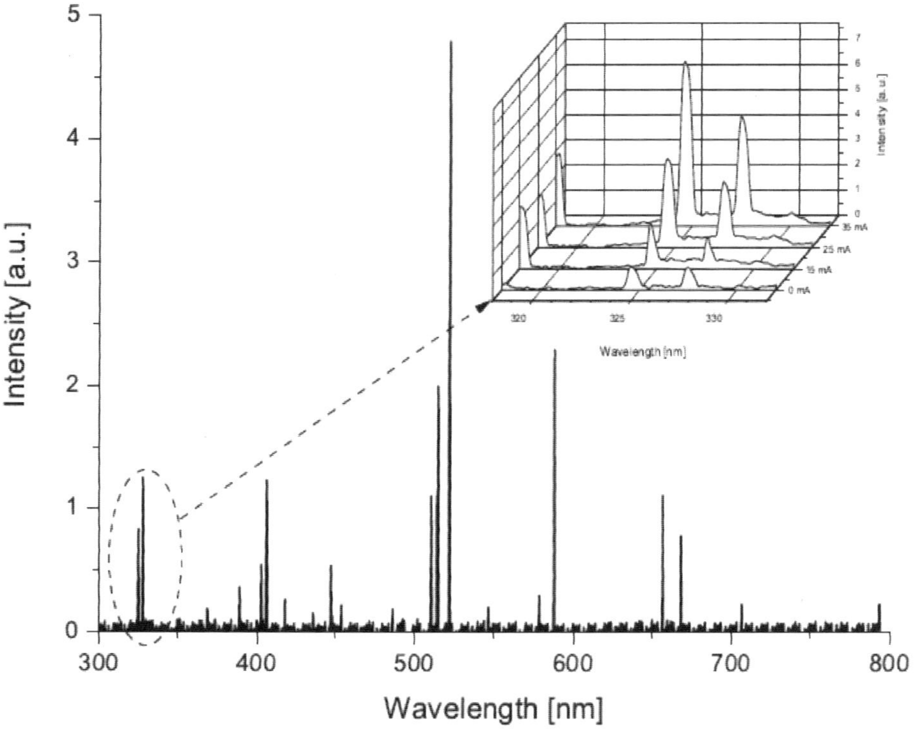

Figure 2. The full spectrum obtained by LIBS on a copper sample. The subfigure shows the intensity of two analytical coper lines (324.75 nm and 327.4 nm) at four different values of the discharge current.

In this spectrum the characteristic lines of copper - 324.75 nm, 327.4 nm, 510.55 nm, 515.32 nm, 522.01 nm, 578.21 nm and those of helium - 447.15 nm, 587.56 nm, 667.81 nm and 706.52 nm are found. A section of this spectrum, from 318 to 332 nm, is plotted in the sub-figure against the spectra obtained from LA-HCD technique at discharge currents of 15, 25 and 35 mA. This plot allows observing the behavior of the copper analytical lines 324.75 and 327.4 nm as a function of discharge current. The stand-alone LIBS plot is marked as 0 mA and the others are accordingly 15, 25 and 35 mA .

The sub-figure shows a well-pronounced enhancement of intensity of the copper analytical lines with discharge current. As it is seen both copper analytical lines increase about five times when changing the discharge current from 15 mA to 35 mA. So the analytical line intensity obtained by LA-HCD is proportional to the value of the discharge current and the enhancement of the intensity is proportional to the discharge input power. Such a behavior of the analytical lines is explained by taking into account the additional excitation of the ablated copper atoms which they undergo in the HCD. This excitation is due to the larger number of excitation collisions between the ablated atoms and the discharge species - electrons, excited atoms and ions. Hence the ablated atoms gain more energy in these collisions and dissipate more energy in radiation processes causing higher intensity of the corresponding analytical lines. This shows that under identical conditions the LA-HCD technique can provide better excitation efficiency than the stand-alone LIBS. This results in more intense analytical signal which leads to better sensitivity.

4. Conclusions

A novel design of a LA-HCD technique is presented. The new features can be observed due to the slit along the HC that allows registration of the longitudinal distribution of the optical signal and its decoupling from the laser excitation. Preliminary results of the modified LA-HCD scheme are presented. Comparison between these results and results from stand-alone LIBS is done. Enhanced analytical performance of the LA-HCD technique is observed. This improvement is attributed to more efficient excitation of the ablated atoms in the HCD. The future work on the modified LA-HCD technique includes: additional optimization of the operational parameters, recording of the temporal evolution of axial species distribution and use of buffer gases other than helium - argon, neon and mixtures of them.

Acknowledgments

The work is supported by Bulgarian Science Fund projects DO 02-274/2008 and DMU 03-14 (NEMO). It is partially sponsored by the EU Catrene project Solution for Energy Efficient Lighting (SEEL).

References

[1] Hahn D, Omenetto N 2010 *Appl. Spectr.* **64** 335A-366A
[2] Windom B and Hahn D 2009 *J. Anal. At. Spectr.* **24** 1665-75
[3] Marcus K 1993 *Glow Discharge Spectroscopies* (New York: Plenum Press) p 423
[4] Iida Y 1990 *Spectr. Chem. Acta B: At. Spectr.* **45** 427-38
[5] Matsuta H, Naeem T and Wagatsuma K 2003 *The Jap. Soc. for Anal. Chem.* **19** 813-14
[6] Naeem T, Matsuta H and Wagatsuma K 2004 *Anal. Bioanal. Chem.* **379** 115-119
[7] Naeem T, Matsuta H and Wagatsuma K 2004 *The Jap. Soc. for Anal. Chem* **20** 1717-1720
[8] van Dijk J, Peerenboom K, Jimenez M, Mihailova D and van der Mullen J 2009 *J. Phys.* D **42** 19402
[9] Mihailova D, Grozeva M, Hagelaar G, van Dijk J, Brok W and van der Mullen J 2008 *J. Phys.* D **41** 24502
[10] Mihailova D, van Dijk J, Hagelaar G, Karatodorov S, Zahariev P, Grozeva M and van der Mullen J 2012 *J. Phys. D: Appl. Phys.* **45** 165201

Collisional electron spectroscopy method for gas analysis

M S Stefanova[1], P M Pramatarov[1], A A Kudryavtsev[2], R A Peyeva[1] and T B Patrikov[1]

[1]Institute of Solid State Physics, Bulgarian Academy of Sciences, 72 Tzarigradsko Chaussee blvd. 1784 Sofia, Bulgaria

[2]St.Petersburg State University, 7-9 Universitetskaya nab, 199034 St.Petersburg, Russia

E-mail: mstef@issp.bas.bg

Abstract. Recently developed collisional electron spectroscopy (CES) method, based on identification of gas impurities by registration of groups of nonlocal fast electrons released by Penning ionization of the impurity particles by helium metastable atoms, is verified experimentally. Detection and identification of atoms and molecules of gas impurities in helium at pressures of 14 - 90 Torr with small admixtures of Ar, Kr, CO2, and N2 are carried out. The nonlocal negative glow plasma of short dc microdischarge is used as most suitable medium. Records of the energy spectra of penning electrons are performed by means of an additional electrode - sensor, located at the boundary of the discharge volume. Maxima appear in the electron energy spectra at the characteristic energies corresponding to Penning ionization of the impurity particles by helium metastable atoms.

1. Introduction

The plasma electron spectroscopy (PLES) method, developed by Kolokolov *at al.* [1] represents a new approach for studies of elementary plasma processes and atomic constants. On the base of the PLES method a new method for identification of gas impurities is proposed by Kudryavtsev [2]. It is based on identification of gas impurities by registration of groups of nonlocal fast electrons, released in reaction of Penning ionization of the impurity particles by helium metastable atoms. Further development of PLES method for identification of unknown species in a gas phase is reported in [3]. In [1-3] the nonlocal afterglow plasma of positive column at low pressures (0.5 – 3.5 Torr) is used, and temporal resolution of the registration system is applied, which complicates the devices.

In this paper the new method of collisional electron spectroscopy (CES), whose basic principles are patented by Kudryavtsev *at al.* [4] is experimentally veryfied. The CES method is based on measurements of energy spectra $R(\varepsilon)$ of fast electrons released in Penning ionization of the impurities atoms or molecules by metastable atoms of the main gas:

$$A^* + B \rightarrow A + B^+ + e\{\varepsilon_p\} \quad (1)$$

The energy of penning electrons is $\varepsilon_p = \varepsilon_m - \varepsilon_i$, where, ε_m is the excitation energy of metastable particles A^* and ε_i is the ionization energy of particles B. When EEDF is formed in nonlocal regime [5], i.e. electron energy relaxation length λ_ε is greater than the characteristic discharge dimension L, $\lambda_\varepsilon > L$, the electrons move in discharge volume with conservation of their full energy $\varepsilon = w + e\phi(r)$ (kinetic plus potential). If the energy of penning electrons is $\varepsilon_p > \Phi_w$ (where Φ_w is the potential between axis and discharge wall), these electrons reach the wall in a free diffusion. In the absence of electric field or when it has a small value (e.g., negative glow plasma, beam plasma), where T_e, and hence $e\Phi_w$ are small, the energy spectrum of penning electrons represent sharp peaks near energies ε_p, reproducing the spectrum $R(\varepsilon)$ of reactions of type (1) [1-3]. Using the measured penning electrons spectra $R(\varepsilon)$ and the well known rate constants for Penning reactions one can identify the impurities and determine their concentrations. The CES method enables identification of gas impurities in a main gas in collisional regime of movement of the particles, where the different groups of electrons do not relax in energy by collisions in the volume and behave independently of each other. This opportunity is associated with the fact that in one elastic collision with a helium atom, the electron loses only a small portion δ of its initial kinetic energy ε_p, $\delta < 10^{-4}$. As a result the electron energy relaxation length λ_ε , considerably exceeds its mean free path λ. In particular for helium,$\lambda_\varepsilon /\lambda \approx 70$, and the parameter $\lambda_\varepsilon p$ has value: $\lambda_\varepsilon p = 3 \div 5$ *cm.Torr*, where p is the gas pressure.

It is convenient to use helium metastable atoms as A^* particles in (1), because their high excitation energy (19.8 *eV*, *20.6 eV*) is sufficient to ionize any other atom or molecule (except neon) and to produce characteristic electrons having kinetic energy of several *eV*.

Since, in nonlocal plasma, each group of penning electrons reaches the plasma boundary with its initial energy, the registration of the EEDF is possible by means of an additional electrode - sensor, located at the boundary of the discharge volume. Thus, it is possible to use a large collecting area of the sensor and to enhance significantly the sensitivity of fast electrons registration.

Over the characteristic length L, the EEDF does not depend essentially on the local plasma parameters and it is the same at any point of the plasma. As a result, the nonlocal plasma has an important property: measuring the EEDF at plasma boundary, we get information from the whole plasma volume.

In this paper a new design of microplasma gas analyzer is proposed and experiments are carried out to detect and identify gas admixtures in helium.

2. Experiment

The design of microplasma gas analyzer used to demonstrate the practical realization of the CES method for gas analysis is shown in figure 1. The negative glow plasma of a short dc microdischarge is employed as most suitable medium for nonlocal formation of the EEDF and for creation of a compact gas analyzer. It consists of plane, disk-shaped (4 mm in diameter) cathode and anode, disposed at 1.7 and 4 mm distance. A ring-shaped additional electrode - sensor is placed coaxially between anode and cathode. It is used to collect the electrons under investigation and to record its VA-characteristics. The sensor is made of 0.4 mm molybdenum wire and has 4 mm diameter. The sensor has large collecting area compared to the classical Langmuir probe – it is comparable to that of the anode. The plasma volume is reduced dramatically; it is about 0.02 - 0.05 cm^3, which is more than two orders of magnitude less, compared to the previous works [1-3]. The small size of the plasma volume permits a significant increase in the operating pressures. Experiments are carried out in the negative glow plasma of a short (without a positive column) dc glow discharge. The gap between the cathode and the anode is chosen in such a manner to fulfill the condition for nonlocal formation of the EEDF at the pressures used in this work.

Records of electron energy spectra are carried out in helium with small admixtures of known gas impurities - argon, krypton, carbon dioxide and nitrogen. The total gas pressure is ranged from 14 to 90 Torr and the discharge current is varied from 0.5 to 5 mA. The identification of atoms and molecules of the impurities is performed by registration of the second derivative of the probe VAcharacteristic d^2I/dV^2, which, according to the Druyvesteyn relation [1], is proportional to the EEDF, i.e. to the energy spectra of penning electrons.

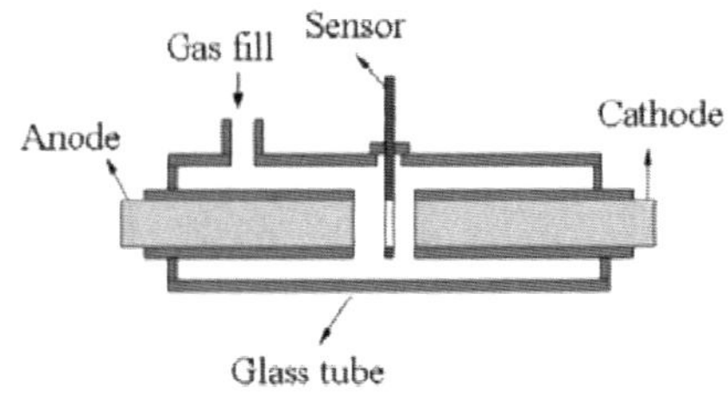

Figure 1. Schematic view of the microplasma gas analyzer.

The "modulation method" [1] is chosen to obtain the second derivative of the sensor current on the applied scanning voltage. Modulating signal of the type $\Delta V = V_0 \cos \omega t$ is used. For that purpose sinusoidal voltage of amplitude $2V_0 = 0.2 - 1.2V$ (peak to peak) and frequency $\omega = 1362$ Hz is mixed with the retarding sensor voltage. Registration system, conventional for probe measurements, is used.

3. Results and discussion

Typical electron energy spectra in helium-krypton gas mixture at two values of the modulation voltage are shown in figure 2 and figure 3. Well expressed maxima are seen. According to equation (1) the maxima recorded in the second derivative are due to electrons released in Penning ionization of krypton atoms by helium metastable atoms. At 1.2 V modulating voltage the characteristic krypton maximum appears at about 6.2 eV (figure 2), while at 0.5 V modulating voltage the characteristic krypton maximum has fine structure (figure 3). Maxima at 5.1 eV, 5.8 eV and 6.6 eV are resolved. The fine structure of the krypton maximum is due to the Penning reactions of triplet $He(2\ ^3S_1)$ and singlet $He(2\ ^1S_0)$ helium metastable atoms with krypton atoms producing krypton ions in two ground states $Kr^+(^2P_{3/2})$ and $Kr^+(^2P_{1/2})$, having energy difference of 0,67 eV. The fine structure components have comparable values. Cross section for Penning ionization of Kr with singlet helium metastable atoms $He(2\ ^1S_0)$ is $28.2.10^{-16}$ cm^2 while for triplet helium metastable atoms $He(2^3S_1)$ it is $7.7.10^{-16}$ cm^2 [6]. Therefore one can conclude for the presence of high enough density of helium singlet metastable atoms in the plasma volume. Besides the krypton maximum at 5.8 eV, another maximum at 19.8 eV is recorded (figure 2). This maximum corresponds to the electrons released in reaction of deactivation of helium metastable atoms $He(2^3S_1)$ by super-elastic collisions with slow electrons. The maximum at 1.8 eV in figure 3 is associated to the presence of residual nitrogen traces in the discharge chamber.

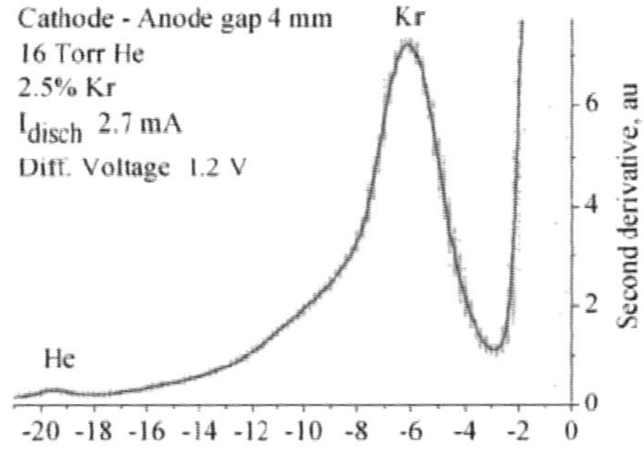

Figure 2. Electron energy spectra in He-Kr mixture.

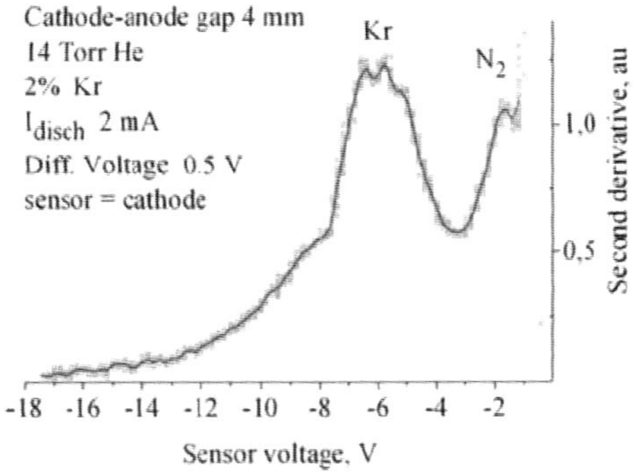

Figure 3. Fine structure of the electron energy spectra in He-Kr mixture.

Hence, the measured energy at the maximum in the EEDF allows to identify the presence of krypton atoms in the plasma.

The electron energy spectrum, when a small portion of argon (1%) is added to 90 Torr helium, is shown in figure 4. Well expressed maximum is seen at 4.1 eV. The characteristic argon maximum is easily obtained at low discharge currents – 2 mA. The argon maximum is due to the Penning reactions of triplet helium metastable atoms with argon atoms.

Figure 5 demonstrates the electron energy spectrum in mixture of 15 Torr helium pressure, 3.3% carbon dioxide and 0.5% nitrogen. Two peaks at 6.8 eV and 4.3 eV are recorded, representing penning electrons released according to reactions of carbon dioxide and nitrogen with singlet and triplet helium metastable atoms:

$$He(2^1S_0) + CO_2(0) \rightarrow He(0) + CO_2^+(X^2\Pi g) + e\ (6{,}81\ eV)$$

$$He(2^3S_1) + N_2(0) \rightarrow He(0) + N_2^+(^2\Sigma^+) + e\ (4{,}24\ eV)$$

It should be noted that the cross section for Penning ionization of CO_2 by $He(2^1S_0)$ is $84.6.10^{-16}$ cm^2, while the cross section of Penning ionization of nitrogen by $He(2^3S_1)$ is $5.2.10^{-16}$ cm^2 [6]. This explains the difference in the measured amplitudes of the two peaks.

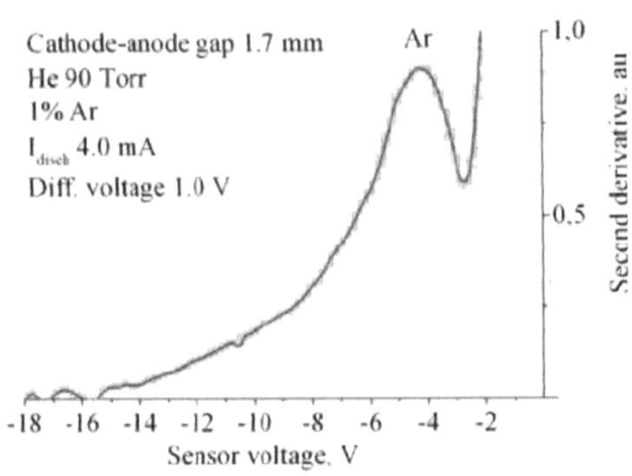

Figure 3. Electron energy spectra in He-Ar mixture.

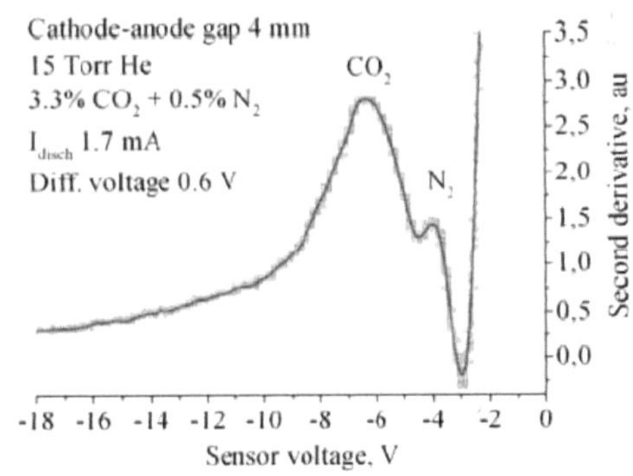

Figure 4. Electron energy spectra in He-CO_2-N_2 mixture.

The recorded maxima in the EEDF demonstrate the possibility to identify the presence of different atomic and molecular admixtures in helium in nonlocal negative glow plasma of a short dc microdischarge at high pressures.

The large sensor surface area collects sufficient number of penning electrons and well expressed maxima in the electron energy spectra are easily obtained at high pressures and low discharge currents. This contributes to significant enhancement in the measurements sensitivity. Temporal resolution of the registration system is not required, which further simplifies the device.

To our knowledge, measurements of maxima in the EEDF at high pressures (14 – 90 Torr) due to electrons produced in Penning reactions of helium metastable atoms with impurity atoms or molecules are not made till now.

4. Conclusions

The CES method gives the opportunity to develop a compact microplasma gas analyser simple in technical performance, whose dimensions are dramatically reduced compared to the traditional methods for gas analysis. All these advantages make it attractive for practical applications.

References

[1] Kolokolov N B, Kudryavtsev A A and Blagoev A B 1994 *Physica Scripta*, **50**, 371
[2] Kudryavtsev A A 1999 A new method of gas Analysis using Penning-electron energy spectra, *Book of abstracts IWRFRI'99, Russian Technologies for Industrial Applications. May 24-26, 1999 , St.Petersburg, Russia, p.94*
[3] Sheverev V A, Khromov N A and. Kojiro D R 2002 *Anal. Chem.* **72**, 5556
[4] Kudryavtsev A A and Tsyganov A B 2007 *US Patent* 7,309,992
[5] Tsendin L D 1995 *Plasma Sources Sci. Technol.* **4** 200
[6] Schmeltekopf A L and Fesenfeld F C 1970 *Jour. Chem. Phys.* **53** 3173

PTFE treatment by remote atmospheric Ar/O_2 plasmas: a simple reaction scheme model proposal

E A D Carbone[1], M W G M Verhoeven[2], W Keuning[3], J J A M van der Mullen[1]

[1] Elementary Processes in Gas discharges (EPG), Department of Applied Physics, Eindhoven University of Technology (TUe), Eindhovenm The Netherlands

[2]Physical Chemistry of Surfaces, Department of Chemical Engineering and Chemistry, Eindhoven University of Technology (TUe), Eindhoven (NL)

[3]Plasma & Materials Processing Group, Dept. of Applied Physics, Eindhoven University of Technology (TUe), Eindhoven (NL)

E-mail: e.a.d.carbone@tue.nl

Abstract. Polytetrafluoroethylene (PTFE) samples were treated by a remote atmospheric pressure microwave plasma torch and analyzed by water contact angle (WCA) and X-ray photoelectron spectroscopy (XPS). In the case of pure argon plasma a decrease of WCA is observed meanwhile an increase of hydrophobicity was observed when some oxygen was added to the discharge. The WCA results are correlated to XPS of reference samples and the change of WCA are attributed to changes in roughness of the samples. A simple kinetics scheme for the chemistry on the PTFE surface is proposed to explain the results.

1. Introduction

Cold atmospheric pressure plasmas (CAPPs) receive nowadays an ever growing interest [1] in many different applications fields like surface treatments, thin film coatings, plasma medicine, waste residues conversion and ozone production.

In order to improve these applications, insight in the physics and chemistry of these plasmas is needed. However, these CAPPs present additional complexity compared to e.g. classical low pressure plasmas. The creation of radicals and active species is no longer ruled by the electrons solely. Instead, many other excitation paths are important that are realized by interactions between excited heavy particles. Moreover molecular assisted recombination processes may also be important in the creation of radicals. In contrast to (atomic) low pressure plasmas the plasma recombination processes do not take place at the walls but mainly in the volume.

This study is devoted to the treatment of polymers using a CAPP in a remote set-up. Since the plasmas are low in gas temperature and far from maxwellian equilibrium the influence of physical processes like thermal heating and ion sputtering can be neglected. These destructive aspects of plasma treatment are further reduced by operating the CAPP in a remote set-up. Remote cool plasma treatments show numerous advantages as they modify [2] only the top layers of the substrate and do not alter the bulk properties of the material.

To probe the kinetics and to understand the action of CAPPs it is important to choose a model-surface with limited reaction paths so that non-ambiguous surface responses can be traced and studied. As a model-surface we selected PTFE (polytetrafluoroethylene), the basic material of teflon. This is a high thermal resistive polymer and has a monomer unit composed only of C and F. Its chemical structure can be defined as $-(CF_2-CF_2)_n-$. This surface is rather non-reactive and exhibits the property of being hydrophobic. Moreover C and F are both easy detectable by means of X-ray photoelectron spectroscopy (XPS). This surface analysis technique can give the stochiometric composition of the 8-9 first monolayers of the polymer substrate and is able to detect all elements of the periodic table apart H and He. With respect to these considerations, PTFE is a natural choice.

The PTFE samples were treated with an atmospheric pressure microwave surfatron and analyzed afterwards by XPS and water contact angle (WCA) measurements. The results are compared with the available literature. A simple reaction scheme model for the surface reactions is eventually proposed to explain the observed behavior.

2. Experimental setup

The plasma is generated by the absorption of electromagnetic waves of 2.45 GHz in an argon flow. The surfatron is a Sairem® launcher coupled to a Microtron™ microwave generator that can deliver powers in the range of 10 to 200W. The Argon and O_2 flow-rates inside the ceramic tube were controlled using Brooks mass flow controllers. An air flow of 20L/min was used as extra cooling of the walls of the plasma tube and the plasma itself. The tube has an inner radius of 0.4 mm and an outer diameter of 1.2 mm. The PTFE samples were mounted on a XY linear motorized translator table in the normal direction to the plasma.

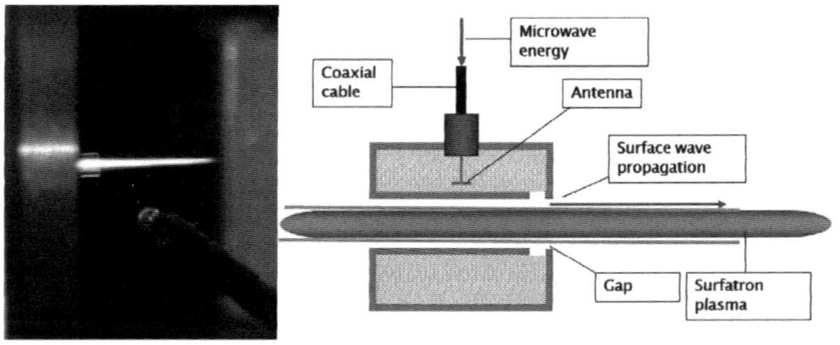

Figure 1. Picture and schematic of the microwave induced plasma (MIP) torch

The PTFE sheets were purchased from Goodfellow (1 mm thick sheet, Ref FP303050). After cutting the sheets to sizes of 1 cm^2, these were cleaned using pure ethanol (Normapur VWR) followed by pure isooctane (GR for analysis, Merck). Subsequently the PTFE samples were treated by the plasma torch. The XPS measurements were carried out with a Kratos AXIS Ultra spectrometer equipped with a monochromatic X-ray source and a delay-line detector (DLD). Spectra were obtained using the aluminium anode (Al Kα) operating at 150W. Survey scans were measured at a constant pass-energy of 160 eV and region-scans at 40 eV. The background pressure was 2×10^{-9} mbar.

The WCA measurements were taken in a static mode with KSV (CAM200) equipment while using distilled 1μl water droplets. The liquid-solid contact angle is a measure of the surface energy of the substrate. For a perfectly flat (at the atomic scale) surface, it can be directly correlated to the composition in chemical groups of the top layers of the sample. However, it also depends on the roughness of the sample. WCA results are then always the convolution of the chemical nature of the surface but also of its morphology.

3. PTFE surface treatment results

The PTFE samples were treated by pure Ar and Ar/O_2 mixtures and their changes in surface properties were probed by WCA. In table 1, it is shown that in the case of pure argon treatment, the hydrophobic character remains nearly constant for short treatment time and that it *decreases* slightly for longer treatments. The lowest WCA-value that was found is 90°. This is a typical value for PTFE after treatment by pure Argon plasma and is usually reported as "cleaning the surface" [10]. For untreated PTFE the WCA values are typically ranged in between 90-110° because they depends strongly of the processing method to manufacture the polymer sheets. It shall be noted that the error on the WCA in these measurements was not more than $\pm 2^o$. Moreover in [3], it was reported that an Argon treatment leads to a slight smoothening of the surface.

Table 1. Water contact angle of PTFE samples treated by the microwave plasma torch: a decrease of the WCA after Argon plasma treatment is observed. In the case of an Ar/O2 plasma treatment (the two bottoms lines) we an increase in WCA and thus the hydrophobicity.

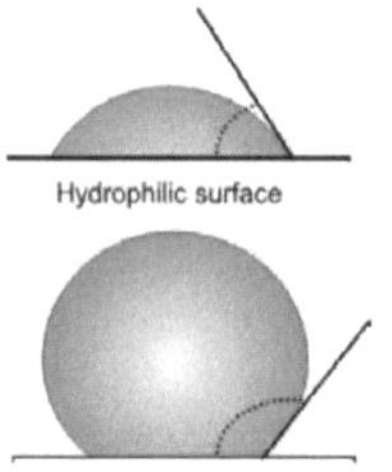

Power (W)	Distance (mm)	Ar/O_2 flux (L-mL/min)	Time (s)	WCA (deg)
Untreated PTFE sample				110
25	2	1.5 - 0	600	90
50	2	1.5 - 0	15	108
50	6	1.5 - 35	120	117
50	4	1.5 - 35	120	126

When some little amount of Oxygen is added (~0.1 %) in the argon discharge, an *increase* of hydrophobicity of the surface is observed contrary to the pure argon case. A higher WCA-value corresponds to a higher degree of hydrophobicity (lower surface energy). Moreover closer distances lead to higher WCA values. This is in contrast to what is usually expected for oxygen plasma treatments where oxygen grafting leads to chemical groups like C-O, C=O or COO on the surface [4]. The polar groups would lead to a reduction the WCA and this is not what we observe.

To get a better understanding of the surface changes, we performed XPS measurements on the plasma-treated samples. The resulting stochiometric findings of the sample surfaces as function of plasma treatment are given in Table 2.

Table 2. Water contact angle of PTFE samples treated by the microwave plasma torch: a decrease of the WCA after Argon plasma treatment is observed. In the case of an Ar/O2 plasma treatment (the two bottoms lines) we an increase in WCA and thus the hydrophobicity.

Power (W)	Distance (mm)	Ar/O_2 flux (L-mL/min)	C (%)	F (%)	O (%)
Untreated PTFE sample			32.10	67.90	0
50	3	3.3 - 35	31.22	67.94	0.84
50	6	3.3 – 35	31.46	67.74	0.80
50	6	3.3 – 17	31.10	68.00	0.90
50	6	3.3 – 0	32.19	66.91	0.90

The table shows that both the Argon and Argon/Oxygen cases do not lead to a change in the C:F ratio. Moreover there is nearly no grafting of oxygen: less than 1%. This is not significant as it is in the error margins of XPS (~1 %).

The deconvolution of the C1s spectra before and after treatment showed no obvious changes in the shape and FWHM of the C1s peak at 292eV. The measured FWHM C1s of PTFE was about 1.2eV before and after the plasma treatments. This indicates that the chemical nature of PTFE – $(CF_2–CF_2)_n$– is conserved in addition to its stochiometry.

These XPS results lead to a straightforward interpretation of the WCA findings. Since there are almost no chemical changes of the surface the decrease of the hydrophobicity of the surface must be attributed to the decrease of roughness in the case of the pure Argon plasma treatment. Conversely, the increase of hydrophobicity in the case of Ar/O_2 plasma can be associated to an increase of roughness of the samples [3, 5].

4. Proposal of a chemical model for the surface kinetics

It is instructive to compare the results of the current study done with a atmospheric *microwave* induced plasma with those obtained by a *RF* atmospheric pressure plasma. In [3], Carbone et al. showed that the addition of oxygen to the RF discharge does not lead to an *increase* but a *decrease* of the surface energy of the PTFE surface. Also in that study, no significant O grafting on the surface was observed by XPS. Moreover the latter indicated towards a conservation of the chemical nature of the polymer. Last but not least, it was found that the roughness of the samples was also increased in the case of Ar/O_2 and this roughening/etching of the surface was confirmed by atomic force microscopy. This is exactly the same kind of behavior that we report here but now for a different plasma source and configuration. Similar trends were also shown in [8,9].

To confirm that the thermal loading of the substrate by means of the plasma cannot explain the changes in trends, we performed Rayleigh scattering on this microwave plasma [19]. The temperature right at the outlet, is found to be as high as xxx K but further downstream decreases down to about 400K. When O_2 is added to the discharge, and using the same technique, it is found that the gas temperature remains constant or even decreases [6] from 750 to 700 K right at the exit of the tube for flow rates from 0 up to 15 sccm in 1 L/min of Ar. The polymers were treated downstream in the region where the plasma was cold (~400K).

The similarity of the results of these two studies, the current one and the one reported in [3] leads to the idea that we face some kind of 'universal' behavior that can be explained by the 'remote' and 'radical chemistry driven' nature of the Ar/O_2 afterglow in contact with surfaces.

To understand this 'universal' behavior a simple reaction scheme model was constructed based on a division of the process in three steps 1) activation, 2) re-association and 3) chemical etching.

4.1. Pure Argon chemistry

The first step, the activation step, leads, in any plasma-surface interactions, to the breaking of some bounds on the surface. This provides the creation of radicals and allows them to react with each other and/or with radicals coming from the plasma. For the PTFE plasma treatment we can define two main roads: one involving the breaking of C-F bounds and the other the breaking of C-C bounds.

4.1.1. An example of C-F bound breaking is the production of negative fluorine ions by electron dissociative attachment [7]:

$$-CF_2-CF_2-CF_2- + e^- \rightarrow -CF_2-{}^\bullet C-F-CF_2- + F^- \quad (1)$$

However in the conditions of this study, defluorination seems negligible as no change in the C:F stochiometry is observed with is its usual signature in fluorinated polymers plasma treatments.

4.1.2. An example of the C-C breaking is the carbon backbone breaking by homolytic (formation of two radicals due to one bound breaking) dissociation. This can be induced by photons, electrons and/or Ar* metastables [8]:

$$-CF_2-CF_2- \; + \; h\nu \, (\text{or } e^-, Ar*) \rightarrow -CF_2^\bullet \; + {}^\bullet CF_2- \; + \; (e^-, Ar) \quad (2)$$

This step will be of main importance for the surface activation and etching as defluorination is negligible, carbon atoms become accessible for surface reactions through C-C bounds breaking.
In the pure Argon plasma case, the recombination of fluorine radicals on the surface can occur [9-10]:

$$-CF_2-(^{\bullet}C-F)-CF_2 + F^{\bullet} \rightarrow -CF_2-\left(F-C-F\right)-CF_2- \quad (3)$$

$$-CF_2-CF_2^{\bullet} + F^{\bullet} \rightarrow -CF_2-CF_3 \quad (4)$$

And more importantly, we will have recombination of fluorocarbon radicals between them directly on the surface [11]:

$$-CF_2-^{\bullet}CF_x + ^{\bullet}CF_y \rightarrow -CF_2-CF_x-CF_y- \quad (5)$$

Although no clear evidence came from our measurements that these processes take place, it cannot be ruled out, moreover it was observed by different authors [3, 9, 14].

4.2. Argon/Oxygen chemistry
The activation steps will be comparable to the pure argon case, meaning that the activation is mainly driven by electrons, photons and Argon metastables rather than by oxygen atoms/molecules. The reason for this assumption is that the ground state O_2 density is only 0.1% of that of argon. However, in the case of this atmospheric pressure plasma jet the dissociation degree of O_2 is unknown but might be expected to be quite high [18]. Nevertheless the emission spectrum of the plasma is still largely characteristic of pure Argon plasma so that we can expect that Argon metastables are still one of the dominant species in the plasma. In the case of Ar/O_2, the main chemistry that occurs is between the radicals formed on the surface with the species coming from the gas phase. The fluorocarbon radicals can either react with atomic oxygen [11] to form *alkoxydes*:

$$-CF_2-(^{\bullet}C-F)-CF_2- + O \rightarrow -CF_2-(F-C-O)-CF_2 \quad (6)$$

or carbonyles [11]:

$$-CF_2-C^{\bullet}-CF_2- + O \rightarrow -CF_2-(C=O)-CF_2- \quad (7)$$

Another reaction path is the direct reaction or the fluorocarbon radicals with molecular oxygen to form *peroxydes* [12-13]:

$$-CF_2-CF_2^{\bullet} + O_2 \rightarrow -CF_2-CF_2O_2^{\bullet} \quad (8)$$

which can also recombine to form *alkoxydes* [13]

$$2(-CF_2-CF_2O_2^{\bullet}) \rightarrow 2(-CF_2-CF_2O^{\bullet}) + O_2 \quad (9)$$

This second path is believed to be of higher importance compared to the steps (6) and (7) because of the large reservoir of O_2 available for reacting on the surface. This is even more likely far in the plasma afterglow where O atoms are expected to recombine to form O_2 and O_3 in the gas phase.

Finally, the intermediates of reaction produced at the steps (6), (7), (8) and (9) can lead to the production of different radicals

$$-CF_2-CF_2-CF_xO_y^{\bullet} + O \rightarrow -CF_2-CF_2^{\bullet} + (CO + CO_2 + CF_20 + CF_2) \quad (10)$$

and other gaseous species as shown in [11-12,15-16] which leave the surface. In that way, they do not contribute anymore to the grafting of C-O$^{\bullet}$, C=O or COO$^{\bullet}$ species on the surface. We have then conservation of the *PTFE-like* character of the polymer after treatment. According to our results, this last step needs to be fast compared to the activation and (partial) oxidation of the carbon atoms. Indeed, this is the only way how, one can expect conservation of the stoihiometric composition of the surface. The chemical etching of the surface will lead ultimately to an increase of roughness of the Teflon sample while oxygen is added to the argon plasma.

5. Conclusions

In this study, we showed that PTFE surfaces can be modified in terms of roughness meanwhile keeping the surface chemistry and stoichiometry nearly unchanged. This behavior seems characteristic of remote Ar/O_2 plasma treatment at atmospheric pressure. A simple three phase chemical scheme is proposed for the interactions between the plasma and the PTFE surface. The first step consists of an activation of the surface by the argon plasma. This is followed by a second step of soft chemical etching of the carbon backbone through radical reactions mainly with molecular oxygen. The fast desorption of the newly formed $C_xF_yO_z$ radicals in the third step allows then finally to obtain no significant changes in the surface stochiometry. Species with a x:y ratio of 0.5 seems to be the ones which leave preferentially the surface with variable amount of O.

Although a complete and exact description of the plasma/surface kinetics would require a full 2D modeling and to implement a complete chemistry model for the Ar/O_2 plasma and its interactions with CF_2-CF_2, a simple approach based on conservation laws was preferred here. The authors would like to point out here the uphill task to find accurate rate coefficients for the surface reactions. They depends strongly of diffusion of chains which depends itself of the length of the polymer chains which is not a conserved parameter during the etching. This simple approach could be extended for other reactive mixtures like Ar/N_2 and Ar/CO_2 which are of main importance in surface treatment and especially for biomedicine applications [17].

References

[1] Journal of "*Plasma Processes and Polymers*", Ed. Wiley, New York
[2] R. d'Agostino, P. Favia, F. Fracassi, Plasma-Processing of Polymers, eds., Kluwer Academic Publisher, NATO ASI Series, E: Applied Sciences, (1997) vol. 346
[3] E. A. D. Carbone, N. Boucher, M. Sferrazza, F. Reniers, SIA (2010) 42, 1014-1018
[4] Jiangnan Lai et al. Applied Surface Science (2006) 252, 3375–3379
[5] A. B. D. Cassie, S. Baxter, Trans. Faraday Soc. (1944) 40, 546
[6] E.A.D. Carbone et al. Proceedings of WELTPP-12 (2009) O8
[7] C.C. Perry, A. J. Wagner, D. H. Fairbrother, Chemical Physics (2002) 280 111–118
[8] S. Trigwell, D. Boucher, C. I. Calle Journal of Electrostatics (2007) 65, 7, 401-407
[9] M. Mona, Occhiello, F. Garbassi, Surf. Interface Anal. (1990) 16, 412417
[10] G. Nang, E. Papirer, P. Fioux, F. Moguet , A. Tressaud, Carbon (1997) 35, 4, 515-528
[11] N. Vandencasteele, B. Nisol, P. Viville, R. Lazzaroni, D. G. Castner, F. Reniers, Plasma Process. Polym. (2008) 5, 7, 661
[12] A. Faucitano, A. Buttafava, S. Karolczak, P.A. Guarda, G. Marchionni, Journal of Fluorine Chemistry (2004) 125 221–241
[13] R.Mishra, S.P.Tripathy, K.K. Dwivedi, D.T. Khathing, S.Ghosh, M.Muller, D.Fink, Radiation Measurements (2003) 37 247–251
[14] D. J. Wilson, A. J. Eccles, T. A. Steele, R. L. Williams, R. C. Pond, Surf. Interface Anal. (2000) 30, 36 – 39
[15] A. Oshima, T. Seguchi, Y. Tabata, Radiation Physics and Chemistry (1999) 55, 61-71
[16] S. R. Allayarov , T. A. Konovalova , A. Waterfield , A. Ligia Focsan , V. Jackson, R. Craciun, L. D. Kispert, J. S. Thrasher, D. A. Dixon, Journal of Fluorine Chemistry (2006) 127, 1294–1301
[17] D. G. Castner , B. D. Ratner Surface Science (2002) 500, 28–60
[18] Kinga Kutasi, Vasco Guerra and Paulo Sá *J. Phys. D: Appl. Phys.(2010)* 43, 175201
[19] A F H van Gessel, E A D Carbone, P J Bruggeman and J J A M van der Mullen *Plasma Sources Sci. Technol.(2012)* 21, 015003

Plasmachemical Conservation of Corroded Metallic Objects

Vera Sazavska, Radka Balastikova, Frantisek Krcma, Lucie Radkova, Petra Fojtikova, Radek Prikryl, Michal Prochazka

Faculty of Chemistry, Brno University of Technology, Purkyňova 118, 612 00 Brno, Czech Republic

Email: xcsazavska@fch.vutbr.cz

Abstract: Plasmachemical process for conservation of metallic objects is a new way of effective and fast treatment of corroded objects. This process consists of two main steps: corrosion removal and deposition of a protecting film. Removal of corrosion products is based on plasmachemical reduction of corrosion layers by radio-frequency (RF) low pressure hydrogen plasma. Chosen barrier films are parylene (poly-para-xylylene) coatings and SiO_2-like high density films. Parylene coatings are prepared by a standard chemical vapor deposition (CVD) method. Plasma enhanced chemical vapour deposition (PECVD) enables preparation of SiO_x based thin films with higher flexibility due to variable incorporated organics groups. The coatings were characterized by various methods in order to obtain information about their chemical structure (FTIR) and barrier properties (OTR). The results from standard corrosion test were compared with those on samples treated by conventional conservation procedures.

1. Introduction

Plasmachemical removal of corrosion layers from metal samples is a relatively new technique for restoration of archaeological artefacts. The method is based on a partial reduction of the incrustation and corrosion layers by RF low pressure hydrogen plasma [1, 2]. The hydrogen molecule is dissociated to very reactive species like radicals, ions, and excited hydrogen in plasma discharge. These species react with compounds from corrosion layers. Corrosion products (like oxides, chlorides, etc.) can be reduced to the pure metal due to the reduction effect of hydrogen.

Surface of the treated object is highly reactive and inclines to rapid oxidation by air oxygen. To prevent this process, it is necessary to protect the surface by a barrier thin film preventing the penetration of oxygen (as well as the other corrosion agents) to the surface. The barrier film must not change optical properties of the object and have to be removable without damage of the object. Suitable barrier films are parylene (poly-para-xylylene) coatings and SiO_2-like high density films.
Parylene coatings are chemically inert, conformal and transparent with excellent barrier properties [3], but relatively small adhesion. Parylene is one of the most well-known chemical vapor deposited (CVD) thin film polymers. SiO_2-like high density films have very good barrier properties and excellent adhesion. SiOx layers were deposited by PECVD in a low pressure reactor with capacitively coupled plasma discharge. Mixture of hexamethyldisiloxane (HMDSO) with oxygen was used as a precursor of plasmachemical reactions. Hexamethyldisiloxane is another suitable monomer for the thin film preparation. The main advantages of these films are good barrier properties. Moreover, they are transparent, colourless, flexible, soft and inert [4].

2. Experimental setup

2.1 Plasmachemical treatment
The experiment was carried out in the Quartz cylindrical reactor (i.d. 95 mm, length of 90 cm) with outer copper electrodes (Figure 1). The capacitive coupled RF power supply (frequency of 13.56 MHz) gave the total power up to 600 W in a continuous or pulsed regime. We used pulses with a duty cycle of 75 %, 50 % and 25 %, and frequency was 1000 Hz (25% pulse means 0.25 ms discharge on and 0.75 ms off). Flowing plasma was created in pure hydrogen (gas flow of 50 sccm) at pressure of 200 Pa. The real temperature of samples was measured continuously by a thermocouple installed inside the model sample. The optical emission spectroscopy (OES) of OH radical was used for the process monitoring.

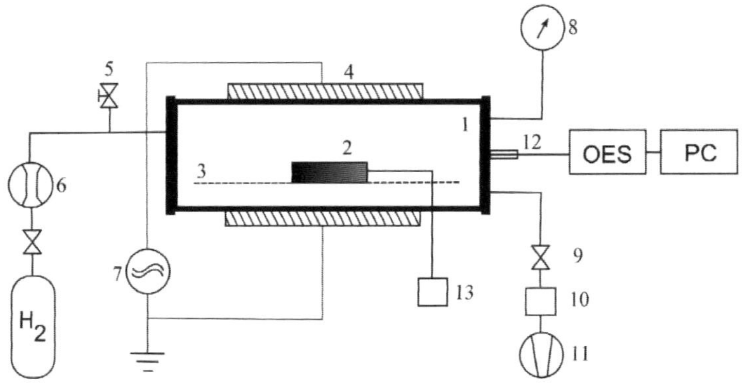

Figure 1: Experimental set-up: 1 – Quartz discharge reactor (90 cm long, i.d. 95 mm); 2 – corroded sample; 3 – glass sample holder; 4 – outer copper electrodes; 5 – air-inlet valve; 6 – mass flow controller; 7 – RF power supply and matching network; 8 – pressure gauge; 9, 10 – valves; 11 – rotary oil pump; 12 – optical fibre; 13 – thermocouple

2.2 Parylene deposition
The apparatus for chemical vapour deposition uses the Gorham process consisted of the following parts: vaporizer, pyrolytic chamber, cooling section and deposition chamber. The vaporizer consists of a brass cartridge where the defined amount of Parylene C is placed. The pyrolytic chamber is about 1.5 m long and its inner diameter is 38 mm. The cooling chamber is shorter than the pyrolytic chamber and also has the bigger inner diameter. The deposition chamber is a vertical cylinder of diameter of 35 cm and a height of approximately 50 cm. Vacuum inside the deposition chamber is up to 10 Pa and it is produced by a mechanical vacuum pump. The last main part of the apparatus is a cold trap with temperature about – 20 °C which serves for removal of the rest of monomer to prevent clogging of the mechanical vacuum pump.

Defined amount of Parylene C (42 g for the barrier thickness of 20 μm) was placed into the vaporizer. Temperature of the vaporizer was gradually increased up to about 120 °C when the process began and further to 180 °C when process was finished. Because of the vacuum gradient, molecules of the dimer went through the pyrolytic chamber which was heated to temperature of about 680 °C. The dimer was fragmented into reactive monomer particles. Monomer particles came into the cooling section which served for the gas temperature decrease (150 °C). Pressure in the deposition chamber was kept constant at the value of 10 Pa. The whole process of the deposition lasted for 4.5 hour.

2.2 pp-HMDSO deposition
The apparatus for SiOx depositions is designed as a high-vacuum bell jar reactor with pressure of about 10^{-4} Pa. The system consists of a glass reactor with volume of 30 L. It is equipped by two thermal evaporators and a capacitive coupled system of electrodes connected to the plasma generator (working frequency of 13.56 MHz). The system also allows monomer vapour deposition and it is equipped with mass flow controllers. Reactor space is evacuated by a rotary oil pump in the first step,

and by a turbomolecular pump, subsequently. It is possible to change the pumping velocity of the whole pumping system continuously by a manually controlled gate valve (DN100) in order to achieve the desired working pressure. The radiofrequency capacitive coupled glow discharge is created between two plan-parallel electrodes with the diameter of 170 mm which are placed in the centre of the reactor. The inlet of working gases is located under the lower grounded electrode. The upper electrode can be settled by substrates (Quartz polished on both sides, silicon wafer, a foil or another deposited sample). Radiofrequency voltage is supplied to this electrode from the generator with an automatic matching network.

The deposition process was done on a silicon wafer and a PP foil as a substrate. The PP foil was cleaned by ethanol. Liquid hexamethyldisiloxan ($C_6H_{18}OSi_2$) was used for plasma polymerization as a precursor. It was placed in a glass flask and only the monomer vapour was let flowing into the apparatus. Pressure inside the deposition chamber was $2 \cdot 10^{-3}$ Pa. Then, the chamber was purged with the mixture of the monomer (0.54 sccm) and oxygen (10 sccm). The deposition time was 6 minutes for all samples and power was 50 W.

3. Results and discussion

3.1 Integral intensity of OH radicals

The optical emission spectroscopy of created OH radicals was used for the process monitoring. Atomic hydrogen reacts with oxygen from corrosion layers and forms OH radical (in excited state) [5]. OH radical emits in the spectral region of 305–325 nm, and its integral intensity was used for the quantitative analysis of oxygen removal from the corrosion layer [6, 7]. Simultaneously, rotational temperature was calculated from emission spectra. The rotational temperature was more or less constant during the plasma treatment (600 K).

Integral intensity of OH radicals during plasma removal of corroded objects is shown in Figure 2. OH intensity quickly increases to the maximum then the reduction process slow down. Plasma treatment was stopped when the value of relative intensity of OH radicals reached one tenth of maximum OH radical intensity [7]. Maximum OH radical intensity shows extent of reduction of oxides from corrosion layers. We can see in the Figure 2 maximum OH radical intensity is lower in pulsed regime (200W 75% pulse) thus reduction process is less strong in the pulsed regime than in continual regime. Advantage of pulsed regime is lower heating stress of treated objects. Pulsed regime is suitable for metal objects sensitive to the heating stress. The plasma treatment duration was 60–120 minutes depending on the duty cycle.

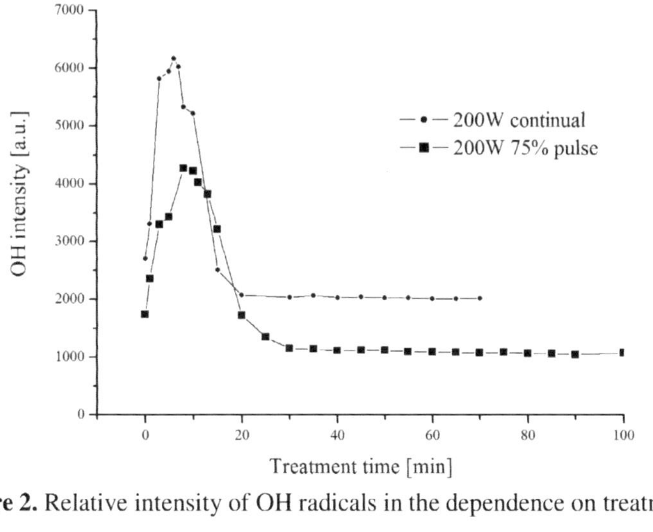

Figure 2. Relative intensity of OH radicals in the dependence on treatment time

Journal of Physics: Conference Series **715** (2016) 012012 doi:10.1088/1742-6596/715/1/012012

3.2 Energy dispersive X-ray spectroscopy of treated samples

Corroded and treated objects were analyzed by a SEM-EDX method in order to determine changes of their surface elemental composition caused by hydrogen plasma. We can see decrease of amount of elements from corrosion layer and increase of abundance of elements from metals (in Figure 3). Thus partial removal of corrosion layers was proved.

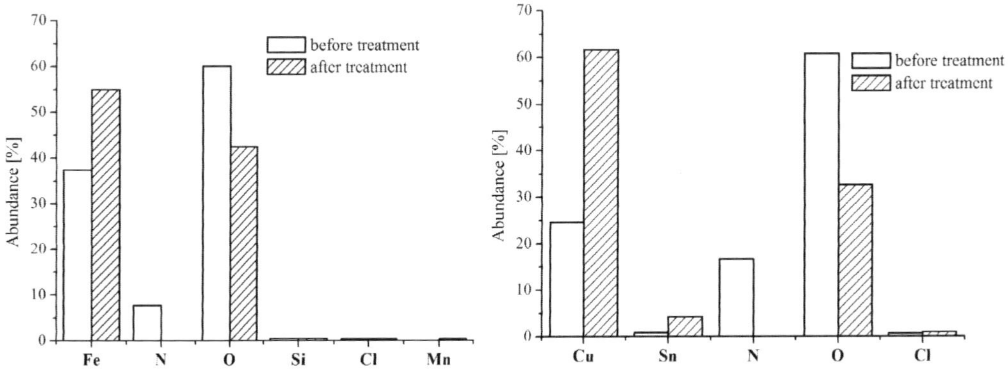

Figure 3: Results of the SEM-EDX analysis of iron (left) and bronze (right) samples

3.3 Fourier Transform Infrared Spectroscopy

Chemical structure analysis of protective films was done by a spectrometer Nicolet iS10 with OMNIC software. The FT-IR measurements give information about the coating composition. The obtained spectrum of Parylene C is in Figure 4 and the spectrum of pp-HMDSO is in Figure 5.

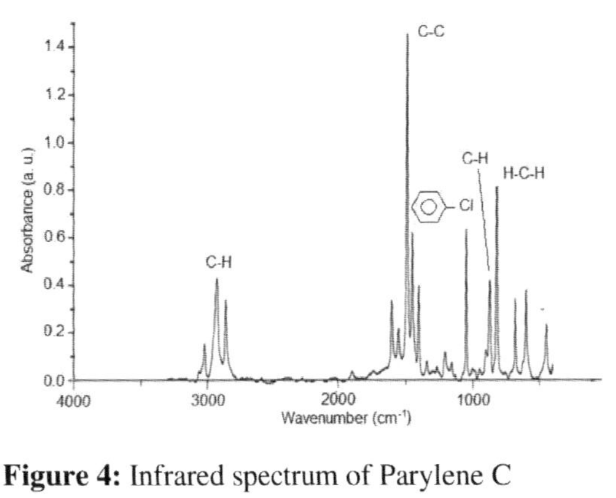

Figure 4: Infrared spectrum of Parylene C

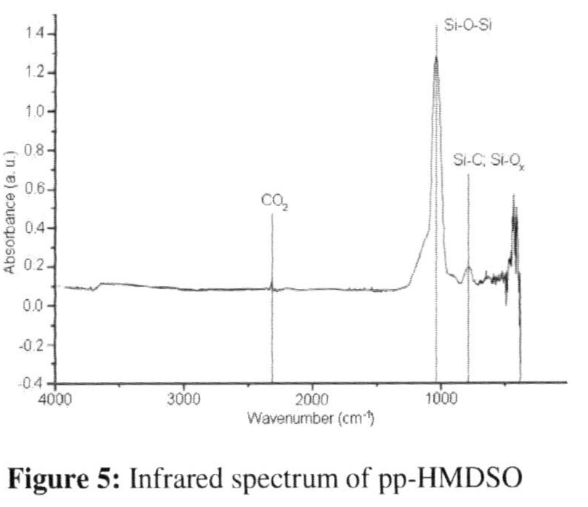

Figure 5: Infrared spectrum of pp-HMDSO

In the spectrum of Parylene C, the 1494 cm^{-1} peak represents the aryl carbon vibrational movement (C – C). Peak at 1050 cm^{-1} is attributed to the chlorine bonding to the ring. The 825 cm^{-1} peak is a representative of two neighbouring hydrogen atoms bonded to the ring (CH_2), and the peak at 877 cm^{-1} is a representative of the single hydrogen to the ring (CH) which is neighbouring chlorine and ethyl group in this case. Additional characteristic peak of CH_2 on the benzene ring is at 1452 cm^{-1}. A pair of peaks at 2861 and 2927 cm^{-1} indicates the vibrational movement of the C–H bond in methyl groups and C–H bond vibrational movement in the aromatic ring at 3019 cm^{-1} [8].

There are smaller peaks in addition to the peaks described in the spectrum. These peaks can be caused by noise or various contaminants contained in the deposition precursor of Parylene C or added substances, such as 3–(trimethoxysilyl)–propyl methacrylate.

In the spectrum of the pp-HMDSO, there is the peak at 2330 cm^{-1} which represents the molecule of CO_2. Peak at 1069 cm^{-1} is a representative of Si-O-Si, and the peak at 790 cm^{-1} represents $Si\text{-}O_x$ bonds [9]. It means that we have prepared a layer with minimum amount of carbon.

3.3 Oxygen Transmission Rate

Transmission rate of the PP foil for oxygen was measured. We have measured foils with the thin layer (Parylene C or pp-HMDSO) as well as clean PP foils. As you can see in Table 1, the values for Oxygen Transmission Rate measurements are very different. In case of pp-HMDSO, the transmission rate is of the same order as the clean PP-foil. It is in contrast to Parylene C where the transmission rate is about 20 times lower than the permeability of the PP-foil.

Table 1: Transmission rate of the PP-foil

Substrate	Permeability ($cm^3/(m^2 \cdot atm \cdot day)$)
PP-foil	665
PP-foil with pp-HMDSO	127
PP-foil with Parylene C	32

3.3 Corrosion test in salt spray chamber

The new protective film (Parylen C) and a barrier film (paraloid B44) commonly used in restoration practice were compared. The metal samples with both films were tested in the salt spray chamber Ascot 450 according to the CSN EN ISO 9227 norm. The tested samples were exposed to the salt spray for 300 h. The salt spray had contained 5% NaCl solution mixed with air in the nozzle. The temperature was 25 °C in the salt spray chamber.

The result of the corrosion test for paraloid B44 is in Figure 6. The protective varnish paraloid B44 have not perfect resistance to corrosion. The Parylen C layer is shown in Figure 7. Only few corrosive spots are presented on the surface. The result is clear that the new Parylen C layer has much better corrosion resistance than commonly used varnish paraloid B44.

Figure 6. Results of the test CSN EN ISO 9227 in the salt chamber for the iron sample with the protective film paraloid B44 (**a** before test, **b** after 300h in the salt chamber)

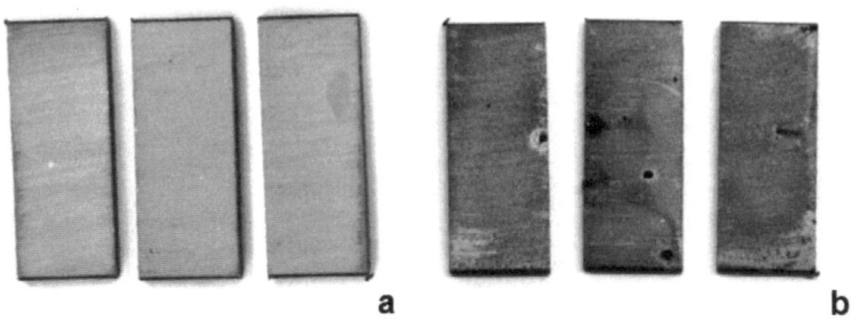

Figure 7. Results of the test CSN EN ISO 9227 in the salt chamber for the iron sample with the protective film Parylen C (**a** before test, **b** after 300h in the salt chamber)

4. Conclusion

We proved partial remove of corrosion compounds by SEM-EDX and OES. We characterised thin protective films (Parylen C and SiO_x) by FTIR and OTR. Elemental composition was investigated by the infrared spectroscopy. We determined that thin layers of pp-HMDSO did not contain carbon, which indicated that we had prepared the thin layer really similar to the SiO_2 like layer. Oxygen permeability through the protective films was investigated by oxygen transmission rate. The thin film of Parylene C had a four times smaller permeability than the thin film of pp-HMDSO.

Parylen C layer was compared with the conventional varnish paraloid B44 by the corrosion test in the salt spray chamber. The corrosion test proved much better protective properties of Parylen C.

Acknowledgements:
This research has been supported by the Czech Ministry of Culture, project No. DF11P01OVV004.

References:

[1] S. Veprek, J. Patscheider, J. Elmer, *Plasma Chem. Plasma Process.* **5** (1985) 201-209
[2] S. Veprek, J. Elmer, Ch. Eckmann, M. Jurcik-Rajman, *J. Electrochem. Soc.* **134** (1987) 2398
[3] E. Meng, P.-Y. Li, Y.-C. Tai, *J. Micromech. Microeng.*, **18** (2008) 045004 (13pp)
[4] S. Zanini, C. Riccardi, M. Orlandi, P. Esena, M. Tontini, M. Milani, V. Cassio, *Surface & Coatings Technology.*, 200 (2005) 953-957
[5] V. Sázavská, L. Řádková, F. Krčma, M. Zmrzlý, R. Přikryl, M. Zahoran:, *Proceedings of 18th Symposium on Application of Plasma Processes,* 266-271 (2011)
[6] Z. Raskova, F. Krcma, M. Klima, J. Kousal, *Czech. J. Phys.* 52 (2002) D927.
[7] V.Sazavska, F. Krcma, T. Simsova, N. Zemanek, *Journal of Physics: Conference Series* **207**, Article No. 012011, 4 pages (2010), ISSN: 1742-6596
[8] R. R. A. Callahan, K. G. Pruden, G. B.Raupp, S. P. Beaudoin, *Journal of Vacuum Science.*, **21**(4) (2003) 1496-1500 ISSN 0734211
[9] A. Patelli, , S. Vezzu, L. Zottarel, E. Menin, , C. Sada, A. Martucci, , S. Costacurta, *Plasma Processes and Polymers.* **6** (2009) 665-670 ISSN 16128850

Institute of Physics
Dirac House, Temple Back
Bristol BS1 6BE UK

ISSN: 1742-6588
ISBN 978-1-5108-2419-5

CURRAN ASSOCIATES INC.
proceedings.com

Eurotherm Seminar No 108 – Nanoscale and Microscale Heat Transfer V

Journal of Physics: Conference Series Volume 785

Santorini, Greece
26 – 30 September 2016

IOP
Institute of Physics